Learning SQL

Other resources from O'Reilly

oreilly.com

oreilly.com is more than a complete catalog of O'Reilly books. You'll also find links to news, events, articles, weblogs, sample chapters, and code examples.

oreillynet.com is the essential portal for developers interested in open and emerging technologies, including new platforms, programming languages, and operating systems.

Conferences

O'Reilly brings diverse innovators together to nurture the ideas that spark revolutionary industries. We specialize in documenting the latest tools and systems, translating the innovator's knowledge into useful skills for those in the trenches. Visit *conferences.oreilly.com* for our upcoming events.

Safari Bookshelf (*safari.oreilly.com*) is the premier online reference library for programmers and IT professionals. Conduct searches across more than 1,000 books. Subscribers can zero in on answers to time-critical questions in a matter of seconds. Read the books on your Bookshelf from cover to cover or simply flip to the page you need. Try it today with a free trial.

Preface

Programming languages come and go constantly, and very few languages in use today have roots going back more than a decade or so. Some examples are Cobol, which is still used quite heavily in mainframe environments, and C, which is still quite popular for operating system and server development and for embedded systems. In the database arena, we have SQL, whose roots go all the way back to the 1970s.

SQL is the language for generating, manipulating, and retrieving data from a relational database. One of the reasons for the popularity of relational databases is that properly designed relational databases can handle huge amounts of data. When working with large data sets, SQL is akin to one of those snazzy digital cameras with the high-power zoom lens in that you can use SQL to look at large sets of data, or you can zoom in on individual rows (or anywhere in between). Other database management systems tend to break down under heavy loads because their focus is too narrow (the zoom lens is stuck on maximum), which is why attempts to dethrone relational databases and SQL have largely failed. Therefore, even though SQL is an old language, it is going to be around for a lot longer and has a bright future in store.

Why Learn SQL?

If you are going to work with a relational database, whether you are writing applications, performing administrative tasks, or generating reports, you will need to know how to interact with the data in your database. Even if you are using a tool that generates SQL for you, such as a reporting tool, there may be times when you need to bypass the automatic generation feature and write your own SQL statements.

Learning SQL has the added benefit of forcing you to confront and understand the data structures used to store information about your organization. As you become comfortable with the tables in your database, you may find yourself proposing modifications or additions to your database schema.

Why Use This Book to Do It?

The SQL language is broken into several categories. Statements used to create database objects (tables, indexes, constraints, etc.) are collectively known as SQL *schema statements*. The statements used to create, manipulate, and retrieve the data stored in a database are known as the SQL *data statements*. If you are an administrator, you will be using both SQL schema and SQL data statements. If you are a programmer or report writer, you may only need to use (or be *allowed* to use) SQL data statements. While this book demonstrates many of the SQL schema statements, the main focus of this book is on programming features.

With only a handful of commands, the SQL data statements look deceptively simple. In my opinion, many of the available SQL books help to foster this notion by only skimming the surface of what is possible with the language. However, if you are going to work with SQL, it behooves you to understand fully the capabilities of the language and how different features can be combined to produce powerful results. I feel that this is the only book that provides detailed coverage of the SQL language without the added benefit of doubling as a "door stop" (you know, those 1,250-page "complete references" that tend to gather dust on people's cubicle shelves).

While the examples in this book run on MySQL, Oracle Database, and SQL Server, I had to pick one of those products to host my sample database and to format the result sets returned by the example queries. Of the three, I chose MySQL because it is freely obtainable, easy to install, and simple to administer. For those readers using a different server, I ask that you download and install MySQL and load the sample database so that you can run the examples and experiment with the data.

Structure of This Book

This book is divided into 13 chapters and 4 appendixes:

- Chapter 1, *A Little Background*, explores the history of computerized databases, including the rise of the relational model and the SQL language.
- Chapter 2, *Creating and Populating a Database*, demonstrates how to create a MySQL database, create the tables used for the examples in this book, and populate the tables with data.
- Chapter 3, *Query Primer*, introduces the select statement and further demonstrates the most common clauses (select, from, where).
- Chapter 4, *Filtering*, demonstrates the different types of conditions that can be used in the where clause of a select, update, or delete statement.
- Chapter 5, *Querying Multiple Tables*, shows how queries can utilize multiple tables via table joins.

- Chapter 6, *Working with Sets*, is all about data sets and how they can interact within queries.

- Chapter 7, *Data Generation, Conversion, and Manipulation*, demonstrates several built-in functions used for manipulating or converting data.

- Chapter 8, *Grouping and Aggregates*, shows how data can be aggregated.

- Chapter 9, *Subqueries*, introduces the subquery (a personal favorite) and shows how and where they can be utilized.

- Chapter 10, *Joins Revisited*, further explores the various types of table joins.

- Chapter 11, *Conditional Logic*, explores how conditional logic (i.e., if-then-else) can be utilized in `select`, `insert`, `update`, and `delete` statements.

- Chapter 12, *Transactions*, introduces transactions and shows how to use them.

- Chapter 13, *Indexes and Constraints*, explores indexes and constraints.

- Appendix A, *ER Diagram for Example Database*, shows the database schema used for all examples in the book.

- Appendix B, *MySQL Extensions to the SQL Language*, demonstrates some of the interesting non-ANSI features of MySQL's SQL implementation.

- Appendix C, *Solutions to Exercises*, shows solutions to the chapter exercises.

- Appendix D, *Further Resources*, suggests where to turn for more advanced training.

Conventions Used in This Book

The following typographical conventions are used in this book:

Italic
> Used for filenames, directory names, and URLs. Also used for emphasis and to indicate the first use of a technical term.

`Constant width`
> Used for code examples and to indicate SQL keywords within text.

`Constant width italic`
> Used to indicate user-defined terms.

UPPERCASE
> Used to indicate SQL keywords within example code.

`Constant width bold`
> Indicates user input in examples showing an interaction. Also indicates emphasized code elements to which you should pay particular attention.

 Indicates a tip, suggestion, or general note. For example, I use notes to point you to useful new features in Oracle9*i*.

Indicates a warning or caution. For example, I'll tell you if a certain SQL clause might have unintended consequences if not used carefully.

How to Contact Us

Please address comments and questions concerning this book to the publisher:

O'Reilly Media, Inc.
1005 Gravenstein Highway North
Sebastopol, CA 95472
(800) 998-9938 (in the United States or Canada)
(707) 829-0515 (international or local)
(707) 829-0104 (fax)

O'Reilly maintains a web page for this book, which lists errata, examples, and any additional information. You can access this page at:

http://www.oreilly.com/catalog/learningsql

To comment or ask technical questions about this book, send email to:

bookquestions@oreilly.com

For more information about O'Reilly books, conferences, Resource Centers, and the O'Reilly Network, see the web site at:

http://www.oreilly.com

Using Code Examples

This book is here to help you get your job done. In general, you may use the code in this book in your programs and documentation. You do not need to contact us for permission unless you're reproducing a significant portion of the code. For example, writing a program that uses several chunks of code from this book does not require permission. Selling or distributing a CD-ROM of examples from O'Reilly books *does* require permission. Answering a question by citing this book and quoting example code does not require permission. Incorporating a significant amount of example code from this book into your product's documentation *does* require permission.

We appreciate, but do not require, attribution. An attribution usually includes the title, author, publisher, and ISBN. For example: "*Learning SQL* by Alan Beaulieu. Copyright 2005 O'Reilly Media, Inc., 0-596-00727-2."

If you feel your use of code examples falls outside fair use or the permission given above, feel free to contact us at *permissions@oreilly.com*.

Safari Enabled

 When you see a Safari® Enabled icon on the cover of your favorite technology book, that means the book is available online through the O'Reilly Network Safari Bookshelf.

Safari offers a solution that's better than e-books. It's a virtual library that lets you easily search thousands of top tech books, cut and paste code samples, download chapters, and find quick answers when you need the most accurate, current information. Try it for free at *http://safari.oreilly.com*.

Acknowledgments

A book is a living thing, and what you now hold in your hands is a far cry from my initial ramblings. The person most responsible for this metamorphosis is my editor, Jonathan Gennick; thank you for your assistance in every step of this project, both for your editorial prowess and your expertise with the SQL language. Next, I would like to acknowledge my three technical reviewers, Peter Gulutzan, Joseph Molinaro, and Jeff Cox, who challenged me to make this book both technically sound and appropriate for readers new to SQL. Also, many thanks to the multitude of people at O'Reilly Media who have helped make this book a reality, including my production editor, Matt Hutchinson; the cover designer, Ellie Volckhausen; and the illustrator, Rob Romano.

CHAPTER 1

A Little Background

Before we roll up our sleeves and get to work, it might be beneficial to introduce some basic database concepts and look at the history of computerized data storage and retrieval.

Introduction to Databases

A *database* is nothing more than a set of related information. A telephone book, for example, is a database of the names, phone numbers, and addresses of all people living in a particular region. While a telephone book is certainly a ubiquitous and frequently used database, it suffers from the following:

- Finding a person's telephone number can be time consuming, especially if the telephone book contains a large number of entries.

- A telephone book is only indexed by last/first names, so finding the names of the people living at a particular address, while possible in theory, is not a practical use for this database.

- From the moment the telephone book is printed, the information becomes less and less accurate as people move into or out of a region, change their telephone numbers, or move to another location within the same region.

The same drawbacks attributed to telephone books can also apply to any manual data storage system, such as patient records stored in a filing cabinet. Because of the cumbersome nature of paper databases, some of the first computer applications developed were *database systems*, which are computerized data storage and retrieval mechanisms. Because a database system stores data electronically rather than on paper, a database system is able to retrieve data more quickly, index data in multiple ways, and deliver up-to-the-minute information to its user community.

Early database systems managed data stored on magnetic tapes. Because there were generally far more tapes than tape readers, technicians were tasked with loading and unloading tapes as specific data was required. Because the computers of that era had very little memory, multiple requests for the same data generally required the data to be read from the tape multiple times. While these database systems were a significant improvement over paper databases, they are a far cry from what is possible with today's technology. (Modern database systems can manage terabytes of data spread across many fast-access disk drives, holding tens of gigabytes of that data in high-speed memory, but I'm getting a bit ahead of myself.)

Nonrelational Database Systems

For the first several decades of computerized database systems, data was stored and represented to users in various ways. In a *hierarchical database system*, for example, data is represented as one or more tree structures. Figure 1-1 shows how data relating to George Blake's and Sue Smith's bank accounts might be represented via tree structures.

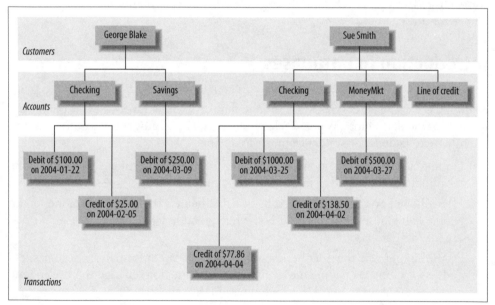

Figure 1-1. Hierarchical view of account data

George and Sue each have their own tree containing their accounts and the transactions on those accounts. The hierarchical database system provides tools for locating a particular customer's tree and then traversing the tree to find the desired accounts and/or transactions. Each node in the tree may have either zero or one parent and zero, one, or many children. This configuration is known as a *single-parent hierarchy*.

Another common approach, called the *network database system*, exposes sets of records and sets of links that define relationships between different records. Figure 1-2 shows how George's and Sue's same accounts might look in such a system.

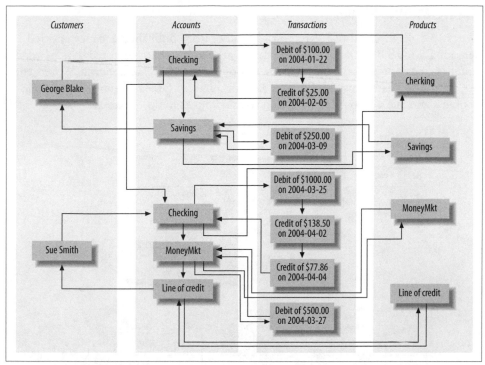

Figure 1-2. Network view of account data

In order to find the transactions posted to Sue's money market account, you would need to perform the following steps:

1. Find the customer record for Sue Smith.
2. Follow the link from Sue Smith's customer record to her list of accounts.
3. Traverse the chain of accounts until you find the money market account.
4. Follow the link from the money market record to its list of transactions.

One interesting feature of network database systems is demonstrated by the set of product records on the far right of Figure 1-2. Notice that each product record (Checking, Savings, etc.) points to a list of account records that are of that product type. Account records, therefore, can be accessed from multiple places (both customer records and product records), allowing a network database to act as a *multiparent hierarchy*.

Both hierarchical and network database systems are alive and well today, although generally in the mainframe world. Additionally, hierarchical database systems have enjoyed a rebirth in the directory services realm, such as Microsoft's Active Directory and Netscape's Directory Server, as well as with Extensible Markup Language (XML). Beginning in the 1970s, however, a new way of representing data began to take root, one that was more rigorous yet easy to understand and implement.

The Relational Model

In 1970 Dr. E. F. Codd of IBM's research laboratory published a paper entitled "A Relational Model of Data for Large Shared Data Banks" that suggested data be represented as sets of *tables*. Rather than using pointers to navigate between related entities, redundant data is used to link records in different tables. Figure 1-3 shows how George's and Sue's account information would appear in this context.

Customer

cust_id	fname	lname
1	George	Blake
2	Sue	Smith

Account

account_id	product_cd	cust_id	balance
103	CHK	1	$75.00
104	SAV	1	$250.00
105	CHK	2	$783.64
106	MM	2	$500.00
107	LOC	2	0

Product

product_cd	name
CHK	Checking
SAV	Savings
MM	Money market
LOC	Line of credit

Transaction

txn_id	txn_type_cd	account_id	amount	date
978	DBT	103	$100.00	2004-01-22
979	CDT	103	$25.00	2004-02-05
980	DBT	104	$250.00	2004-03-09
981	DBT	105	$1000.00	2004-03-25
982	CDT	105	$138.50	2004-04-02
983	CDT	105	$77.86	2004-04-04
984	DBT	106	$500.00	2004-03-27

Figure 1-3. Relational view of account data

There are four tables in Figure 1-3 representing the four entities discussed so far: customer, product, account, and transaction. Looking across the customer table in Figure 1-3, you can see three *columns*: cust_id (which contains the customer's ID number), fname (which contains the customer's first name), and lname (which contains the customer's last name). Looking down the customer table, you can see two *rows*, one containing George Blake's data and the other containing Sue Smith's data. The number of columns that a table may contain differs from server to server, but it is generally large enough not to be an issue (Microsoft SQL Server, for example, allows up to 1,024 columns per table). The number of rows that a table may contain is more a matter of physical limits (i.e., how much disk drive space is available) than of database server limitations.

Each table in a relational database includes information that uniquely identifies a row in that table (known as the *primary key*), along with additional information needed to describe the entity completely. Looking again at the customer table, the cust_id column holds a different number for each customer; George Blake, for example, can be uniquely identified by customer ID #1. No other customer will ever be assigned that identifier, and no other information is needed to locate George Blake's data in the customer table. While I might have chosen to use the combination of the fname and lname columns as the primary key (a primary key consisting of two or more columns is known as a *compound key*), there could easily be two or more people with the same first and last names that have accounts at the bank. Therefore, I chose to include the cust_id column in the customer table specifically for use as a primary-key column.

Some of the tables also include information used to navigate to another table. For example, the account table includes a column called cust_id, which contains the unique identifier of the customer who opened the account, along with a column called product_cd, which contains the unique identifier of the product to which the account will conform. These columns are known as *foreign keys*, and they serve the same purpose as the lines that connect the entities in the hierarchical and network versions of the account information. However, unlike the rigid structure of the hierarchical/network models, relational tables can be used in various ways (including some not envisioned by the people who originally designed the database).

It might seem wasteful to store the same data many times, but the relational model is quite clear on what redundant data may be stored. For example, it is proper for the account table to include a column for the unique identifier of the customer who opened the account, but it is not proper to include the customer's first and last names as well. If a customer were to change her name, for example, you want to make sure that there is only one place in the database that holds the customer's name; otherwise, the data might be changed in one place but not another, causing the data in the database to be unreliable. The proper place for this data is the customer table, and only the cust_id data should be included in other tables. It is also not proper for a single column to contain multiple pieces of information, such as a name column that contains both a person's first and last names, or an address column that contains street, city, state, and zip code information. The process of refining a database design to ensure that each independent piece of information is in only one place (except for foreign keys) is known as *normalization*.

Getting back to the four tables in Figure 1-3, you may wonder how you would use these tables to find George Blake's transactions against his checking account. First, you would find George Blake's unique identifier in the customer table. Then, you would find the row in the account table whose cust_id column contains George's unique identifier and whose product_cd column matches the row in the product table whose name column equals "checking." Finally, you would locate the rows in the transaction table whose account_id column matches the unique identifier from the account table. This might sound complicated, but it can be done in a single command using the SQL language as you will see shortly.

Some Terminology

I've introduced some new terminology in the previous sections, so maybe it's time for some formal definitions. Table 1-1 shows the terms we will use for the remainder of the book along with their definitions.

Table 1-1. Terms and definitions

Term	Definition
Entity	Something of interest to the database user community. Examples include customers, parts, geographic locations, etc.
Column	An individual piece of data stored in a table.
Row	A set of columns that together completely describe an entity or some action on an entity. Also called a record.
Table	A set of rows, held either in memory (nonpersistent) or on permanent storage (persistent).
Result set	Another name for a nonpersistent table, generally the result of an SQL query.
Primary key	One or more columns that can be used as a unique identifier for each row in a table.
Foreign key	One or more columns that can be used together to identify a single row in another table.

What Is SQL?

Along with Codd's definition of the relational model, he proposed a language called DSL/Alpha for manipulating the data in relational tables. Shortly after Codd's paper was released, IBM commissioned a group to build a prototype based on Codd's ideas. This group created a simplified version of DSL/Alpha that they called SQUARE. Refinements to SQUARE led to a language called SEQUEL, which was, finally, renamed SQL.

SQL is now entering its fourth decade, and it has undergone a great deal of change along the way. In the mid 1980s, the American National Standards Institute (ANSI) began working on the first standard for the SQL language, which was published in 1986. Subsequent refinements led to new releases of the SQL standard in 1989, 1992, 1999, and 2003. Along with refinements to the core language, new features have been added to the SQL language to incorporate object-oriented functionality, among other things.

SQL goes hand-in-hand with the relational model because the result of an SQL query is a table (also called, in this context, a *result set*). Thus, a new permanent table can be created in a relational database simply by storing the result set of a query. Similarly, a query can use both permanent tables and the result sets from other queries as inputs (this will be explored in detail in Chapter 9).

One final note: SQL is not an acronym for anything (although many people will insist it stands for "Structured Query Language"). When referring to the language, it is equally acceptable to say the letters individually (i.e., S. Q. L.) or to use the word "sequel."

SQL Statement Classes

The SQL language is broken into several distinct parts: the parts that will be explored in this book include SQL schema statements, which are used to define the data structures stored in the database; SQL data statements, which are used to manipulate the data structures previously defined using SQL schema statements; and SQL transaction statements, which are used to begin, end, and rollback transactions (covered in Chapter 12). For example, to create a new table in your database, you would use the SQL schema statement create table, whereas the process of populating your new table with data would require the SQL data statement insert.

To give you a taste of what these statements look like, here's an SQL schema statement that creates a table called corporation:

```
CREATE TABLE corporation
 (corp_id SMALLINT,
  name VARCHAR(30),
  CONSTRAINT pk_corporation PRIMARY KEY (corp_id)
 );
```

This statement creates a table with two columns, corp_id and name, with the corp_id column identified as the primary key for the table. The finer details of this statement, such as the different data types available with MySQL, will be probed in the following chapter. Next, here's a SQL data statement that inserts a row into the corporation table for Acme Paper Corporation:

```
INSERT INTO corporation (corp_id, name)
VALUES (27, 'Acme Paper Corporation');
```

This statement adds a row to the corporation table with a value of 27 for the corp_id column and a value of Acme Paper Corporation for the name column.

Finally, here's a simple select statement to retrieve the data that was just created:

```
mysql< SELECT name
    -> FROM corporation
    -> WHERE corp_id = 27;
+------------------------+
| name                   |
+------------------------+
| Acme Paper Corporation |
+------------------------+
```

All database elements created via SQL schema statements are stored in a special set of tables called the *data dictionary*. This "data about the database" is known collectively as *metadata*. Just like tables that you create yourself, data dictionary tables can be queried via a select statement, thereby allowing you to discover the current data structures deployed in the database at runtime. For example, if you are asked to write a report showing the new accounts created last month, you could either hard-code the names of the columns in the account table that were known to you when you wrote the report, or you could query the data dictionary to determine the current set of columns and dynamically generate the report each time it is executed.

Most of this book will be concerned with the data portion of the SQL language, which consists of the select, update, insert, and delete commands. SQL schema statements will be demonstrated in Chapter 2, where the sample database used throughout this book will be generated. In general, SQL schema statements do not require much discussion apart from their syntax, whereas SQL data statements, while few in number, offer numerous opportunities for detailed study. Most chapters in this book will, therefore, concentrate on the SQL data statements.

SQL: A Nonprocedural Language

If you have worked with programming languages in the past, you are used to defining variables and data structures, using conditional logic (i.e., if-then-else) and looping constructs (i.e., do while ... end), and breaking your code into small, reusable pieces (i.e., objects, functions, procedures). Your code is handed to a compiler, and the resulting executable does exactly (well, not always *exactly*) what you programmed it to do. Whether you work with Java, C#, C, Visual Basic, or some other *procedural* language, you are in complete control of what the program does. With SQL, however, you will need to give up some of the control you are used to, because SQL statements define the necessary inputs and outputs, but the manner in which a statement is executed is left up to a component of your database engine known as the *optimizer*. The optimizer's job is to look at your SQL statements and, taking into account how your tables are configured and what indexes are available, decide the most efficient execution path (well, not always the *most* efficient). Most database engines will allow you to influence the optimizer's decisions by specifying *optimizer hints*, such as suggesting that a particular index be used; most SQL users, however, will never get to this level of sophistication and will leave such tweaking to their database administrator or performance expert.

With SQL, therefore, you will not be able to write complete applications. Unless you are writing a simple script to manipulate certain data, you will need to integrate SQL with your favorite programming language. Some database vendors have done this for you, such as Oracle with their PL/SQL language or Microsoft with their Trans-actSQL language. With these languages, the SQL data statements are part of the language's grammar, allowing you to seamlessly integrate database queries with procedural commands. If you are using a non-database-specific language such as Java, however, you will need to use a toolkit to execute SQL statements from your code. Some of these toolkits are provided by your database vendor, whereas others are created by third-party vendors or by open-source providers. Table 1-2 shows some of the available options for integrating SQL into a specific language.

Table 1-2. SQL integration toolkits

Language	Toolkit
Java	JDBC (Java Database Connectivity) (JavaSoft)
C++	RogueWave SourcePro DB (third-party tool to connect to Oracle, SQL Server, MySQL, Informix, DB2, Sybase, and PostgreSQL databases)
C/C++	Pro*C (Oracle) MySQL C API (open source) DB2 Call Level Interface (IBM)
C#	ADO.NET (Microsoft)
VisualBasic	ADO.NET (Microsoft)

If you only need to execute SQL commands interactively, every database vendor provides at least a simple tool for submitting SQL commands to the database engine and inspecting the results. Most vendors provide a graphical tool as well that includes one window showing your SQL commands and another window showing the results from your SQL commands. Since the examples in this book are executed against a MySQL database, I will be using the *mysql* command-line tool to run the examples and format the results.

SQL Examples

Earlier in this chapter, I promised to show you an SQL statement that would return all of the transactions against George Blake's checking account. Without further ado, here it is:

```
SELECT t.txn_id, t.txn_type_cd, t.date, t.amount
FROM customer c INNER JOIN account a ON c.cust_id = a.cust_id
  INNER JOIN product p ON p.product_cd = a.product_cd
  INNER JOIN transaction t ON t.account_id = a.account_id
WHERE c.fname = 'George' and c.lname = 'Blake'
  AND p.name = 'checking';
```

Without going into too much detail at this point, this query identifies the row in the account table for George Blake and the row in the product table for the "checking" product, finds the row in the account table for this customer/product combination, and returns four columns from the transaction table for all transactions posted to this account. I will cover all of the concepts in this query (plus a lot more) in the following chapters, but I wanted to at least show what the query would look like.

The previous query contains three different *clauses*: select, from, and where. Almost every query that you encounter will include at least these three clauses, although there are several more that can be used for more specialized purposes. The role of each of these three clauses is demonstrated by the following:

```
SELECT /* one or more things */ ...
FROM /* one or more places */ ...
WHERE /* one or more conditions apply */ ...
```

 Most SQL implementations treat any text between the /* and */ tags as comments.

When constructing your query, your first task is generally to determine which table or tables will be needed and then add them to your from clause. Next, you will need to filter out the data from these tables that doesn't help answer your query and add these conditions to your where clause. Finally, you will decide which columns from the different tables need to be retrieved and add them to your select clause. Here's a simple example that shows how you would find all customers with the last name "Smith":

```
SELECT cust_id, fname
FROM customer
WHERE lname = 'Smith'
```

This query searches the customer table for all rows whose lname column matches the string "Smith" and returns the cust_id and fname columns from those rows.

Along with querying your database, you will most likely be involved with populating and modifying the data in your database. Here's a simple example of how you would insert a new row into the product table:

```
INSERT INTO product (product_cd, name)
VALUES ('CD', 'Certificate of Depysit')
```

Whoops, look like you misspelled "Deposit." No problem. You can clean that up with an update statement:

```
UPDATE product
SET name = 'Certificate of Deposit'
WHERE product_cd = 'CD';
```

Notice that the update statement also contains a where clause, just like the select statement. This is because an update statement must isolate the rows to be modified; in this case, you are specifying that only those rows whose product_cd column matches the string "CD" should be modified. Since the product_cd column is the primary key for the product table, you should expect your update statement to modify exactly one row (or zero, if the value doesn't exist in the table). Whenever you execute an SQL data statement, you will receive feedback from the database engine as to how many rows were affected by your statement. If you are using an interactive tool such as the *mysql* command-line tool mentioned earlier, then you will receive feedback concerning how many rows were either:

- Returned by your select statement
- Created by your insert statement
- Modified by your update statement
- Removed by your delete statement

If you are using a procedural language with one of the toolkits mentioned earlier, the toolkit will include a call to ask for this information after your SQL data statement has executed. In general, it's a good idea to check this info to make sure your statement didn't do something unexpected (like when you forget to put a where clause on your delete statement and delete every row in the table!).

What Is MySQL?

Relational databases have been available commercially for over two decades. Some of the most mature and popular products include:

- Oracle Database from Oracle Corporation
- SQL Server from Microsoft
- DB2 Universal Database from IBM
- Sybase Adaptive Server from Sybase
- Informix Dynamic Server from IBM

All of these database servers do approximately the same thing, although some are better equipped to run very large or very-high-throughput databases. Others are better at handling objects or very large files or XML documents, etc. Additionally, all of these servers do a pretty good job of being compliant with the latest ANSI SQL standard. This is a good thing, and I will make it a point to show you how to write SQL statements that will run on any of these platforms with little or no modification.

Along with the commercial database servers, there has been quite a bit of activity in the open-source community in the past five years with the goal of creating a viable alternative to the commercial database servers. Two of the most commonly used open-source database servers are PostgreSQL and MySQL. The MySQL web site (*http://www.mysql.com*) currently claims over 6 million installations, their server is available for free, and I have found it extremely simple to download and install their server. For these reasons, I have decided that all examples for this book will be run against a MySQL (Version 4.1.11) database, and that the *mysql* command-line tool will be used to format query results. Even if you are already using another server and never plan to use MySQL, I urge you to install the latest MySQL server, load the sample schema and data, and experiment with the data and examples in this book.

However, keep in mind the following caveat:

> This is not a book about MySQL's SQL implementation.

Rather, this book is designed to teach you how to craft SQL statements that will run on MySQL with no modifications, and will run on recent releases of Oracle Database, Sybase Adaptive Server, and SQL Server with little or no modification. If you are using one of the IBM servers mentioned earlier, you might have a bit more work to do.

To keep the code in this book as vendor-independent as possible, I will refrain from demonstrating some of the interesting things that the MySQL SQL language implementers have decided to do that can't be done on other database implementations. Instead, Appendix B covers some of these features for those readers planning to continue using MySQL.

What's in Store

The overall goal of the next four chapters is to introduce the SQL data statements, with a special emphasis of the three main clauses of the `select` statement. Additionally, you will see many examples that use the bank schema (introduced in the next chapter), which will be used for all examples in the book. It is my hope that your growing familiarity with a single database will allow you to get to the crux of an example without your having to stop and examine the tables being used each time.

After you have a solid grasp on the basics, the remaining chapters will drill deep on additional concepts, most of which are independent of each other. Thus, if you find yourself getting confused, you can always move ahead and come back later to revisit a chapter. When you have finished the book and worked through all of the examples, you will be well on your way to becoming a seasoned SQL practitioner.

For those readers interested in learning more about relational databases, the history of computerized database systems, or the SQL language than was covered in this short introduction, here are a few resources worth checking out:

- *Database in Depth: Relational Theory for Practitioners* by C.J. Date (O'Reilly)
- *An Introduction to Database Systems*, Eighth Edition by C.J. Date (Addison Wesley)
- *The Database Relational Model: A Retrospective Review and Analysis: A Historical Account and Assessment of E. F. Codd's Contribution to the Field of Database Technology* by C.J. Date (Addison Wesley)
- *http://en.wikipedia.org/wiki/Database_management_system*
- *http://www.mcjones.org/System_R/*

CHAPTER 2

Creating and Populating a Database

This chapter will provide you with the information you need to create your first database and to create the tables and associated data used for the examples in this book. You will also learn about various data types and see how to create tables using them. Because the examples in this book are executed against a MySQL database, this chapter is somewhat skewed toward MySQL's features and syntax, but most concepts are applicable to any server.

Creating a MySQL Database

If you already have a MySQL database server available for your use, you can start with item number 8 in the instructions below. Keep in mind, however, that this book assumes that you are using MySQL Version 4.1.11 or later, so you may want to consider upgrading your server or installing another server if you are using an earlier release.

The following instructions will show you the minimum steps required to install a MySQL server on a Windows computer, create a database, and load the sample data for this book:

1. Download the MySQL Database Server (Version 4.1.7 or later) from *http://dev. mysql.com*. Unless you are planning to use the server for more than just a training tool, you should download the Essentials Package, which includes only the commonly used tools, instead of the Complete Package.

2. Launch the installation by double-clicking on the downloaded file.

3. Install the server using the "typical install." The installation should be quick and painless, but feel free to consult the online installation guide at *http://dev.mysql. com/doc/mysql/en/Installing.html*.

4. When the installation is complete, make sure the checkbox is checked next to "Configure the MySQL Server now" before pressing the finish button. This will launch the Configuration Wizard.

5. When the Configuration Wizard launches, choose the Standard Configuration radio button, and then check both the Install as Windows Service and Include Bin Directory in Windows Path checkboxes.

6. During the configuration, you will be asked to choose a password for the root user. Make sure you write down the password for future use.

7. Open a shell (use Start → Run → Command) and log in as the root user: `mysql -u root -p`. You will be asked for the root password, and then the `mysql>` prompt will appear.

8. Create a new user for the book data. I created a user called lrngsql using the command: `grant all privileges on *.* to 'lrngsql'@'localhost' identified by 'xxxxx';` (replace *xxxxx* with the password you have chosen for this user).

9. Quit the session using `quit;` and from the shell, log in as your new user via `mysql -u lrngsql -p`.

10. Create a database. I created a database called "bank" using `create database bank;`.

11. Choose your new database via `use bank;`.

12. Download the sample data for this book. You should find the file in the Examples section for this book at *http://www.oreilly.com/catalog/learningsql*.

13. From the *mysql* command-line tool, load the data from the downloaded file using the `source` command, as in `source c:\tmp\learning_sql.sql`. You should replace the path "c:\tmp\" with the location where you stored the sample data script.

You should now have a working database populated with all of the data needed for the examples in this book.

Using the mysql Command-Line Tool

Whenever you invoke the *mysql* command-line tool, you can specify the username and database to use, as in the following:

```
mysql -u lrngsql -p bank
```

You will be asked for your password, and then the `mysql>` prompt will appear, via which you will be able to issue SQL statements and view the results. For example, if you want to know the current date and time, you could issue the following query:

```
mysql> SELECT now( );
+---------------------+
| now( )              |
+---------------------+
| 2005-05-06 16:48:46 |
+---------------------+
1 row in set (0.01 sec)
```

The now( ) function is a built-in MySQL function that returns the current date and time. As you can see, the *mysql* command-line tool formats the results of your queries within a rectangle bounded by +, -, and | characters. After the results have been exhausted (in this case, there is only a single row of results), the *mysql* command-line tool shows how many rows were returned and how long the SQL statement took to execute.

About Missing from Clauses

With some database servers, you won't be able to issue a query without a from clause that names at least one table. Oracle Database is a commonly used server for which this is true. For cases when you only need to call a function, Oracle provides a table called dual, which consists of a single column called dummy that contains a single row of data. In order to be compatible with Oracle Database, MySQL also provides a dual table. The previous query to determine the current date and time could therefore be written as:

```
mysql> SELECT now( )
    FROM dual;
+---------------------+
| now( )              |
+---------------------+
| 2005-05-06 16:48:46 |
+---------------------+
1 row in set (0.01 sec)
```

If you are not using Oracle and have no need to be compatible with Oracle, you can ignore the dual table altogether.

When you are done with the *mysql* command-line tool, simply type **quit;** or **exit;** to return to the shell.

MySQL Data Types

In general, all of the popular database servers have the capacity to store the same types of data, such as strings, dates, and numbers. Where they typically differ is in the specialty data types, such as XML documents or very large text or binary documents. Since this is an introductory book on SQL, and since 98% of the columns you encounter will be simple data types, this book will concern itself only with the character, date, and numeric data types.

Character Data

Character data can either be stored as fixed-length or variable-length strings, the difference being that fixed-length strings are right-padded with spaces, whereas variable-length strings are not. When defining a character column, you must specify the maximum size of any string to be stored in the column. For example, if you want to store strings up to 20 characters in length, you could use either of the following definitions:

```
CHAR(20)    /* fixed-length */
VARCHAR(20) /* variable-length */
```

The maximum length for these data types is currently 255 characters (although upcoming releases will allow for longer strings). If you need to store longer strings (such as emails, XML documents, etc.), then you will want to use one of the text types (`tinytext`, `text`, `mediumtext`, `longtext`), which are covered later in this section. In general, you should use the char type when all strings to be stored in the column are of the same length, such as state abbreviations, and the varchar type when strings to be stored in the column are of varying lengths. Both char and varchar are used in a similar fashion in all of the major database servers.

 Oracle Database is an exception when it comes to the use of varchar. Oracle users should use the varchar2 type when defining variable-length character columns.

Character sets

For languages that use the Latin alphabet, such as English, there is a sufficiently small number of characters such that only a single byte is needed to store each character. Other languages, such as Japanese and Korean, contain large numbers of characters, thus requiring multiple bytes of storage for each character. Such character sets are therefore called *multibyte character sets*.

MySQL can store data using various character sets, both single- and multibyte. To view the supported character sets in your server, you can use the show command, as in:

```
mysql> SHOW CHARACTER SET;
+----------+-----------------------------+---------------------+---------+
| Charset  | Description                 | Default collation   | Maxlen  |
+----------+-----------------------------+---------------------+---------+
| big5     | Big5 Traditional Chinese    | big5_chinese_ci     |    2    |
| dec8     | DEC West European           | dec8_swedish_ci     |    1    |
| cp850    | DOS West European           | cp850_general_ci    |    1    |
| hp8      | HP West European            | hp8_english_ci      |    1    |
| koi8r    | KOI8-R Relcom Russian       | koi8r_general_ci    |    1    |
| latin1   | ISO 8859-1 West European    | latin1_swedish_ci   |    1    |
| latin2   | ISO 8859-2 Central European | latin2_general_ci   |    1    |
| swe7     | 7bit Swedish                | swe7_swedish_ci     |    1    |
| ascii    | US ASCII                    | ascii_general_ci    |    1    |
| ujis     | EUC-JP Japanese             | ujis_japanese_ci    |    3    |
| sjis     | Shift-JIS Japanese          | sjis_japanese_ci    |    2    |
| hebrew   | ISO 8859-8 Hebrew           | hebrew_general_ci   |    1    |
| tis620   | TIS620 Thai                 | tis620_thai_ci      |    1    |
| euckr    | EUC-KR Korean               | euckr_korean_ci     |    2    |
```

```
| koi8u    | KOI8-U Ukrainian         | koi8u_general_ci    |    1 |
| gb2312   | GB2312 Simplified Chinese | gb2312_chinese_ci  |    2 |
| greek    | ISO 8859-7 Greek         | greek_general_ci    |    1 |
| cp1250   | Windows Central European | cp1250_general_ci   |    1 |
| gbk      | GBK Simplified Chinese   | gbk_chinese_ci      |    2 |
| latin5   | ISO 8859-9 Turkish       | latin5_turkish_ci   |    1 |
| armscii8 | ARMSCII-8 Armenian       | armscii8_general_ci |    1 |
| utf8     | UTF-8 Unicode            | utf8_general_ci     |    3 |
| ucs2     | UCS-2 Unicode            | ucs2_general_ci     |    2 |
| cp866    | DOS Russian              | cp866_general_ci    |    1 |
| keybcs2  | DOS Kamenicky Czech-Slovak | keybcs2_general_ci |    1 |
| macce    | Mac Central European     | macce_general_ci    |    1 |
| macroman | Mac West European        | macroman_general_ci |    1 |
| cp852    | DOS Central European     | cp852_general_ci    |    1 |
| latin7   | ISO 8859-13 Baltic       | latin7_general_ci   |    1 |
| cp1251   | Windows Cyrillic         | cp1251_general_ci   |    1 |
| cp1256   | Windows Arabic           | cp1256_general_ci   |    1 |
| cp1257   | Windows Baltic           | cp1257_general_ci   |    1 |
| binary   | Binary pseudo charset    | binary              |    1 |
| geostd8  | GEOSTD8 Georgian         | geostd8_general_ci  |    1 |
+----------+--------------------------+---------------------+------+
```

When I installed the MySQL server, the latin1 character set was automatically chosen as the default character set. However, you may choose to use a different character set for each character column in your database, and you can even store different character sets within the same table. To choose a character set other than the default when defining a column, simply name one of the supported character sets after the type definition, as in:

```
VARCHAR(20) CHARACTER SET utf8
```

With MySQL, you may also set the default character set for your entire database:

```
CREATE DATABASE foreign_sales CHARACTER SET utf8;
```

While this is as much information regarding character sets as I'm willing to discuss in an introductory book, there is a great deal more to the topic of internationalization than what is shown here. If you plan to deal with multiple or unfamiliar character sets, you may want to pick up a book such as *Java Internationalization* (O'Reilly) or *Unicode Demystified: A Practical Programmer's Guide to the Encoding Standard* (Addison Wesley).

Text data

If you need to store data that might exceed the 255-character limit for char and varchar columns, you will need to use one of the text types.

Table 2-1 shows the available text types and their maximum sizes.

Table 2-1. MySQL text types

Text type	Maximum number of characters
Tinytext	255
Text	65,535
Mediumtext	16,777,215
Longtext	4,294,967,295

When choosing to use one of the text types, you should be aware of the following:

- If the data being loaded into a text column exceeds the maximum size for that type, the data will be truncated.
- Unlike a varchar column, trailing spaces will not be removed when data is loaded into the column.
- When using text columns for sorting or grouping, only the first 1,024 bytes are used, although this limit may be increased if necessary.
- The different text types are unique to MySQL. SQL Server has a single text type for large character data, whereas DB2 and Oracle use a data type called clob, for Character Large Object.

If you are creating a column for free-form data entry, such as a notes column to hold data about customer interactions with your company's customer service department, then you will not want to limit the column to 255 characters and should choose either the text or mediumtext types.

 Oracle Database allows up to 2,000 bytes for char columns and 4,000 bytes for varchar columns. SQL Server can handle up to 8,000 bytes for both char and varchar data. Therefore, you are less likely to need to move to a text data type with Oracle or SQL Server than you are with MySQL. Beginning with Version 5.0.3 (currently in Beta), however, MySQL will leapfrog both servers by allowing up to 65,535 bytes for char and varchar columns.

Numeric Data

Although it might seem reasonable to have a single numeric data type called "numeric," there are actually several different numeric data types that reflect the various ways in which numbers are used, as shown below:

A column indicating whether a customer order has been shipped
This type of column, referred to as a *Boolean*, would contain a 0 to indicate false and a 1 to indicate true.

A system-generated primary key for a transaction table
This data would generally start at 1 and increase in increments of 1 up to a potentially very large number.

An item number for a customer's electronic shopping basket

The values for this type of column would be positive whole numbers between 1 and, at most, 200 (for shopaholics).

Positional data for circuit board drill machine

High-precision scientific or manufacturing data often requires accuracy to eight decimal points.

To handle these types of data (and more), MySQL has several different numeric data types. The most commonly used numeric types are those used to store whole numbers. When specifying one of these types, you may also specify that the data is *unsigned*, which tells the server that all data stored in the column will be greater than zero. Table 2-2 shows the five different data types used to store whole-number integers.

Table 2-2. MySQL integer types

Type	Signed range	Unsigned range
Tinyint	−128 to 127	0 to 255
Smallint	−32,768 to 32,767	0 to 65,535
Mediumint	−8,388,608 to 8,388,607	0 to 16,777,215
Int	−2,147,483,648 to 2,147,483,647	0 to 4,294,967,295
Bigint	−9,223,372,036,854,775,808 to 9,223,372,036,854,775,807	0 to 18,446,744,073,709,551,615

When you create a column using one of the integer types, MySQL will allocate an appropriate amount of space to store the data, which ranges from 1 byte for a tinyint to 8 bytes for a bigint. Therefore, you should try to choose a type that will be large enough to hold the biggest number you can envision being stored in the column without needlessly wasting storage space.

For floating-point numbers (such as 3.1415927), you may choose from the numeric types shown in Table 2-3.

Table 2-3. MySQL floating-point types

Type	Numeric range
Float(p,s)	−3.402823466E+38 to −1.175494351E-38 and 1.175494351E-38 to 3.402823466E+38
Double(p,s)	1.7976931348623157E+308 to −2.2250738585072014E-308 and 2.2250738585072014E-308 to 1.7976931348623157E+308

When using a floating-point type, you can specify a *precision* (the total number of allowable digits both to the left and right of the decimal point) and a *scale* (the number of allowable digits to the right of the decimal point), but they are not required. These values are represented in Table 2-3 as *p* and *s*. If you specify a precision and scale for your floating-point column, remember that the data stored in the column will be rounded if the number of digits exceeds the scale and/or precision of the column. For example, a column defined as float(4,2) will store a total of four digits, two to the left of the decimal and two to the right of the decimal. Therefore, such a column would handle the numbers 27.44 and 8.19 just fine, but the number 17.8675 would be rounded to 17.87, and the number 178.5 would be rounded (severely) to 99.99, which is the largest number that can be stored in the column.

Like the integer types, floating-point columns can be defined as unsigned, but this designation only prevents negative numbers from being stored in the column rather than altering the range of data that may be stored in the column.

Temporal Data

Along with strings and numbers, you will almost certainly be working with information about dates and/or times. This type of data is referred to as *temporal*, and some examples of temporal data in a database include:

- The future date that a particular event is expected to happen, such as shipping a customer's order
- The date that a customer's order was shipped
- The date and time that a user modified a particular row in a table
- An employee's birth date
- The year corresponding to a row in a yearly_sales fact table in a data warehouse
- The elapsed time needed to complete a wiring harness on an automobile assembly line

MySQL includes data types to handle all of these situations. Table 2-4 shows the temporal data types supported by MySQL.

Table 2-4. MySQL temporal types

Type	Default format	Allowable values
Date	YYYY-MM-DD	1000-01-01 to 9999-12-31
Datetime	YYYY-MM-DD HH:MI:SS	1000-01-01 00:00:00 to 9999-12-31 23:59:59
Timestamp	YYYY-MM-DD HH:MI:SS	1970-01-01 00:00:00 to 2037-12-31 23:59:59
Year	YYYY	1901 to 2155
Time	HHH:MI:SS	-838:59:59 to 838:59:59

While database servers store temporal data in various ways, the purpose of a format string (second column of Table 2-4) is to show how the data will be represented when retrieved, along with how a date string should be constructed when inserting or updating a temporal column. Thus, if you wanted to insert the date March 23, 2005, into a date column using the default format YYYY-MM-DD, you would use the string '2005-03-23'. Chapter 7 will fully explore how temporal data is constructed and displayed.

Each database server allows a different range of dates for temporal columns. Oracle Database accepts dates ranging from 4712 BC to 9999 AD, while SQL Server only handles dates ranging from 1753 AD to 9999 AD. Although this might not make any difference for most systems that track current and future events, it is important to keep in mind if you are storing historical dates.

The various components of the date formats shown in Table 2-4 are described in Table 2-5.

Table 2-5. Date format components

Component	Definition	Range
YYYY	Year, including century	1000 to 9999
MM	Month	01 (January) to 12 (December)
DD	Day	01 to 31
HH	Hour	01 to 24
HHH	Hours (elapsed)	−838 to 838
MI	Minute	01 to 60
SS	Second	01 to 60

Here's how the various temporal types would be used to implement the examples shown earlier:

- Columns to hold the expected future shipping date of a customer order and an employee's birth date would use the date type, since it is unnecessary to know at what time a person was born and unrealistic to schedule a future shipment down to the second.

- A column to hold information about when a customer order was actually shipped would use the datetime type, since it is important to track not only the date that the shipment occurred but the time as well.

- A column that tracks when a user last modified a particular row in a table would use the timestamp type. The timestamp type holds the same information as the datetime type (year, month, day, hour, minute, second), but a timestamp column will automatically be populated with the current date/time by the MySQL server when a row is added to a table or when a row is later modified.

- A column holding just year data would use the year type.

- Columns that hold data about the length of time needed to complete a task would use the time type. For this type of data, it would be unnecessary and confusing to store a date component, since you are only interested in the number of hours/minutes/seconds needed to complete the task. This information could be derived using two datetime columns (one for the task start date/time, and the other for the task completion date/time) and subtracting one from the other, but it is simpler to use a single `time` column.

Chapter 7 will explore how to work with each of these temporal data types.

Table Creation

Now that you have a firm grasp on what data types may be stored in a MySQL database, it's time to see how to use these types in table definitions. Let's start by defining a table to hold information about a person.

Step 1: Design

A good way to start designing a table is to do a bit of brainstorming to see what kind of information would be helpful to include. Here's what I came up with after thinking for a short time about the types of information that describe a person:

- Name
- Gender
- Birth date
- Address
- Favorite foods

This is certainly not an exhaustive list, but it's good enough for now. The next step is to assign column names and data types. Table 2-6 shows a first pass at these.

Table 2-6. Person table, first pass

Column	Type	Allowable values
Name	Varchar(40)	
Gender	Char(1)	M, F
Birth_date	Date	
Address	Varchar(100)	
Favorite_foods	Varchar(200)	

The `name`, `address`, and `favorite_foods` columns are of type `varchar` and allow for free-form data entry. The `gender` column allows a single character which should only equal M or F. The `birth_date` column is of type `date`, since a time component is not needed.

Step 2: Refinement

In Chapter 1, you were introduced to the concept of *normalization*, which is the process of ensuring that there are no duplicate (other than foreign keys) or compound columns in your database design. In looking at the person columns a second time, the following issues arise:

- The name column is actually a compound object consisting of a first name and a last name.
- Since multiple people can have the same name, gender, birth date, etc., there are no columns in the person table that guarantee uniqueness.
- The address column is also a compound object consisting of street, city, state/province, country, and postal code.
- The favorite_foods column is a list containing 0, 1, or more independent items. It would be best to create a separate table for this data that includes a foreign key to the person table so you know to which person a particular food may be attributed.

After taking these issues into consideration, Table 2-7 gives a normalized version of the person table.

Table 2-7. Person table, second pass

Column	Type	Allowable values
Person_id	Smallint (unsigned)	
First_name	Varchar(20)	
Last_name	Varchar(20)	
Gender	Char(1)	M, F
Birth_date	Date	
Street	Varchar(30)	
City	Varchar(20)	
State	Varchar(20)	
Country	Varchar(20)	
Postal_code	Varchar(20)	

Now that the person table has a primary key (person_id) to guarantee uniqueness, the next step is to build a favorite_food table that includes a foreign key to the person table. Table 2-8 shows the result.

Table 2-8. Favorite_food table

Column	Type
Person_id	Smallint (unsigned)
Food	Varchar(20)

The person_id and food columns comprise the primary key of the favorite_food table, and the person_id column is also a foreign key to the person table.

Step 3: Building SQL Schema Statements

Now that the design is complete for the two tables holding person information, the next step is to generate SQL statements to create the tables in the database. Here is the statement to create the person table:

```
CREATE TABLE person
 (person_id SMALLINT UNSIGNED,
  fname VARCHAR(20),
  lname VARCHAR(20),
  gender CHAR(1),
  birth_date DATE,
  address VARCHAR(30),
  city VARCHAR(20),
  state VARCHAR(20),
  country VARCHAR(20),
  postal_code VARCHAR(20),
  CONSTRAINT pk_person PRIMARY KEY (person_id)
 );
```

Everything in this statement should be fairly self-explanatory except for the last item; when you define your table, you need to tell the database server what column or columns will serve as the primary key for the table. This is done by creating a *constraint* on the table. There are several types of constraints that can be added to a table definition. This constraint is a *primary-key constraint*. It is placed on the person_id column and given the name pk_person. I am in the habit of prefixing my primary-key constraint names with pk_ and then tacking on the table name so that it is obvious what I am looking at when reviewing a list of such constraints.

While on the topic of constraints, there is another type of constraint that would be useful for the person table. In Table 2-7, I added a third column to show the allowable values for certain columns (such as 'M' and 'F' for the gender column). There is another type of constraint called a *check constraint* that constrains the allowable values for a particular column. MySQL allows a check constraint to be attached to a column definition, as in the following:

```
gender CHAR(1) CHECK (gender IN ('M','F')),
```

While check constraints operate as expected on most database servers, the MySQL server allows check constraints to be defined but does not enforce them. However, MySQL does provide another character data type called enum that merges the check constraint into the data type definition. Here's what it would look like for the gender column definition:

```
gender ENUM('M','F'),
```

Here's how the person table definition looks with enum data types for the gender column:

```
CREATE TABLE person
 (person_id SMALLINT UNSIGNED,
  fname VARCHAR(20),
```

```
    lname VARCHAR(20),
    gender ENUM('M','F'),
    birth_date DATE,
    address VARCHAR(30),
    city VARCHAR(20),
    state VARCHAR(20),
    country VARCHAR(20),
    postal_code VARCHAR(20),
    CONSTRAINT pk_person PRIMARY KEY (person_id)
    );
```

Later in this chapter, you will see what happens if you try to add data to a column that violates its check constraint (or, in the case of MySQL, its enumeration values).

You are now ready to run the create table statement using the *mysql* command-line tool. Here's what it looks like:

```
mysql> CREATE TABLE person
    -> (person_id SMALLINT UNSIGNED,
    ->  fname VARCHAR(20),
    ->  lname VARCHAR(20),
    ->  gender ENUM('M','F'),
    ->  birth_date DATE,
    ->  address VARCHAR(30),
    ->  city VARCHAR(20),
    ->  state VARCHAR(20),
    ->  country VARCHAR(20),
    ->  postal_code VARCHAR(20),
    ->  CONSTRAINT pk_person PRIMARY KEY (person_id)
    -> );
Query OK, 0 rows affected (0.27 sec)
```

After processing the create table statement, the MySQL server returns the message "Query OK, 0 rows affected," which tells me that the statement had no syntax errors. If you want to make sure that the person table does, in fact, exist, you can use the describe command (or desc for short) to look at the table definition:

```
mysql> DESC person;
+-------------+----------------------+------+-----+---------+-------+
| Field       | Type                 | Null | Key | Default | Extra |
+-------------+----------------------+------+-----+---------+-------+
| person_id   | smallint(5) unsigned |      | PRI | 0       |       |
| fname       | varchar(20)          | YES  |     | NULL    |       |
| lname       | varchar(20)          | YES  |     | NULL    |       |
| gender      | enum('M','F')        | YES  |     | NULL    |       |
| birth_date  | date                 | YES  |     | NULL    |       |
| address     | varchar(30)          | YES  |     | NULL    |       |
| city        | varchar(20)          | YES  |     | NULL    |       |
| state       | varchar(20)          | YES  |     | NULL    |       |
| country     | varchar(20)          | YES  |     | NULL    |       |
| postal_code | varchar(20)          | YES  |     | NULL    |       |
+-------------+----------------------+------+-----+---------+-------+
10 rows in set (0.06 sec)
```

Columns 1 and 2 of the describe output are self-explanatory. Column 3 shows whether or not a particular column can be omitted when data is inserted into the table. I purposefully left this topic out of the discussion for now (see the sidebar "What Is Null?" for a short discourse), but it will be explored fully in Chapter 4. The fourth column shows whether or not a column takes part in any keys (primary or foreign); in this case, the person_id column is marked as the primary key. Column 5 shows whether a particular column will be populated with a default value if you omit the column when inserting data into the table. The person_id column shows a default value of 0, although this would work only once, since each row in the person table must contain a unique value for this column (since it is the primary key). The sixth column (called "Extra") shows any other pertinent information that might apply to a column.

What Is Null?

In some cases, it is not possible or applicable to provide a value for a particular column in your table. For example, when adding data about a new customer order, the ship_date column cannot yet be determined. In this case, the column is said to be *null* (note that I do not say that it *equals* null), which indicates the absence of a value.

When designing a table, you may specify which columns are allowed to be null (the default), and which columns are not allowed to be null (designated by adding the keywords not null after the type definition).

Now that you've created the person table, your next step is to create the favorite_food table:

```
mysql> CREATE TABLE favorite_food
    -> (person_id SMALLINT UNSIGNED,
    ->    food VARCHAR(20),
    ->    CONSTRAINT pk_favorite_food PRIMARY KEY (person_id, food),
    ->    CONSTRAINT fk_person_id FOREIGN KEY (person_id)
    ->      REFERENCES person (person_id)
    -> );
Query OK, 0 rows affected (0.10 sec)
```

This should look very similar to the create table statement for the person table, with the following exceptions:

- Since a person can have more than one favorite food (which is the reason this table was created in the first place), it takes more than just the person_id column to guarantee uniqueness in the table. This table, therefore, has a two-column primary key: person_id and food.

- The favorite_food table contains another type of constraint called a *foreign-key constraint*. This constrains the values of the person_id column in the favorite_food table to include *only* values found in the person table. With this constraint in place, I will not be able to add a row to the favorite_food table indicating that person_id 27 likes pizza if there isn't already a row in the person table having person_id of 27.

 If you forget to create the foreign-key constraint when you first create the table, you can add it later via the alter table statement.

Describe shows the following after executing the create table statement:

```
mysql> DESC favorite_food;
+--------------+----------------------+------+-----+---------+-------+
| Field        | Type                 | Null | Key | Default | Extra |
+--------------+----------------------+------+-----+---------+-------+
| person_id    | smallint(5) unsigned |      | PRI | 0       |       |
| food         | varchar(20)          |      | PRI |         |       |
+--------------+----------------------+------+-----+---------+-------+
```

Now that the tables are in place, the next logical step is to add some data.

Populating and Modifying Tables

With the person and favorite_food tables in place, you can now begin to explore the four SQL data statements: insert, update, delete, and select.

Inserting Data

Since there is not yet data in the person or favorite_food tables, the first of the four SQL data statements to be explored will be the insert statement. There are three main components to an insert statement:

- The name of the table into which to add the data
- The names of the columns in the table to be populated
- The values with which to populate the columns

Therefore, you are not required to provide data for every column in the table (unless all the columns in the table have been defined as not null). In some cases, those columns that are not included in the initial insert statement will be given a value later via an update statement. In other cases, a column may never receive a value for a particular row of data (such as a customer order that is cancelled before being shipped, thus rendering the ship_date column inapplicable).

Generating numeric key data

Before inserting data into the person table, it would be useful to discuss how values are generated for numeric primary keys. Other than picking a number out of thin air, you have a couple of options:

- Look at the largest value currently in the table and add 1.
- Let the database server provide the value for you.

While the first option may seem valid, it proves problematic in a multiuser environment, since two users might look at the table at the same time and generate the same value for the primary key. Instead, all database servers on the market today provide a safer, more robust method for generating numeric keys. In some cases, such as the Oracle Database, a separate schema object is used (called a *sequence*); in the case of MySQL, however, you simply need to turn on the *auto-increment* feature for your primary-key column. Normally, you would do this at table creation, but doing it now provides the opportunity to learn another SQL schema statement, one that modifies the definition of an existing table:

```
ALTER TABLE person MODIFY person_id SMALLINT UNSIGNED AUTO_INCREMENT;
```

This statement essentially redefines the person_id column in the person table. If you describe the table, you will now see the auto-increment feature listed under the Extra column for person_id:

```
mysql> DESC person;
+-------------+----------------------+------+-----+---------+----------------+
| Field       | Type                 | Null | Key | Default | Extra          |
+-------------+----------------------+------+-----+---------+----------------+
| person_id   | smallint(5) unsigned |      | PRI | NULL    | auto_increment |
| .           |                      |      |     |         |                |
| .           |                      |      |     |         |                |
| .           |                      |      |     |         |                |
```

When you insert data into the person table, simply provide a null value for the person_id column, and MySQL will populate the column with the next available number (by default, MySQL starts at 1 for auto-increment columns).

The insert statement

Now that all the pieces are in place, it's time to add some data. The following statement creates a row in the person table for William Turner:

```
mysql> INSERT INTO person
    -> (person_id, fname, lname, gender, birth_date)
    -> VALUES (null, 'William','Turner', 'M', '1972-05-27');
Query OK, 1 row affected (0.01 sec)
```

The feedback ("Query OK, 1 row affected") tells you that your statement syntax was proper, and that one row was added to the database (since it was an insert statement). You can look at the data just added to the table by issuing a select statement:

```
mysql> SELECT person_id, fname, lname, birth_date
    -> FROM person;
+-----------+---------+--------+------------+
| person_id | fname   | lname  | birth_date |
+-----------+---------+--------+------------+
|         1 | William | Turner | 1972-05-27 |
+-----------+---------+--------+------------+
1 row in set (0.06 sec)
```

As you can see, the MySQL server generates a value of 1 for the primary key. Since there is only a single row in the person table, I neglected to specify which row I am interested in and simply retrieved all of the rows in the table. If there were more than one row in the table, however, I could add a where clause to specify that I only want to retrieve data for the row having a value of 1 for the person_id column:

```
mysql> SELECT person_id, fname, lname, birth_date
    -> FROM person
    -> WHERE person_id = 1;
+-----------+---------+--------+------------+
| person_id | fname   | lname  | birth_date |
+-----------+---------+--------+------------+
|         1 | William | Turner | 1972-05-27 |
+-----------+---------+--------+------------+
1 row in set (0.00 sec)
```

While this query specifies a particular primary-key value, you can use any column in the table to search for rows, as shown by the following query, which finds all rows with a value of 'Turner' for the lname column:

```
mysql> SELECT person_id, fname, lname, birth_date
    -> FROM person
    -> WHERE lname = 'Turner';
+-----------+---------+--------+------------+
| person_id | fname   | lname  | birth_date |
+-----------+---------+--------+------------+
|         1 | William | Turner | 1972-05-27 |
+-----------+---------+--------+------------+
1 row in set (0.00 sec)
```

Before moving on, there are a couple of things about the earlier insert statement that are worth mentioning:

- Values were not provided for any of the address columns. This is fine, since nulls are allowed for those columns.

- The value provided for the birth_date column was a string. As long as you match the required format shown in Table 2-4, MySQL will convert the string to a date for you.

- The column names and the values provided must correspond in number and type. If you name seven columns and only provide six values, or if you provide values that cannot be converted to the appropriate data type for the corresponding column, you will receive an error.

William has also provided information about his favorite foods, so here are three insert statements to store his food preferences:

```
mysql> INSERT INTO favorite_food (person_id, food)
    -> VALUES (1, 'pizza');
Query OK, 1 row affected (0.01 sec)

mysql> INSERT INTO favorite_food (person_id, food)
    -> VALUES (1, 'cookies');
Query OK, 1 row affected (0.00 sec)

mysql> INSERT INTO favorite_food (person_id, food)
    -> VALUES (1, 'nachos');
Query OK, 1 row affected (0.01 sec)
```

Here's a query that retrieves William's favorite foods in alphabetical order using an order by clause:

```
mysql> SELECT food
    -> FROM favorite_food
    -> WHERE person_id = 1
    -> ORDER BY food;
+---------+
| food    |
+---------+
| cookies |
| nachos  |
| pizza   |
+---------+
3 rows in set (0.02 sec)
```

The order by clause tells the server how to sort the data returned by the query. Without the order by clause, there is no guarantee that the data in the table will be retrieved in any particular order.

So that William doesn't get lonely, you can execute another insert statement to add Susan Smith to the person table:

```
mysql> INSERT INTO person
    -> (person_id, fname, lname, gender, birth_date,
    ->  address, city, state, country, postal_code)
    -> VALUES (null, 'Susan','Smith', 'F', '1975-11-02',
    ->  '23 Maple St.', 'Arlington', 'VA', 'USA', '20220');
Query OK, 1 row affected (0.01 sec)
```

If you query the table again, you will see that Susan's row has been assigned the value 2 for its primary-key value:

```
mysql> SELECT person_id, fname, lname, birth_date
    -> FROM person;
+-----------+---------+--------+------------+
| person_id | fname   | lname  | birth_date |
+-----------+---------+--------+------------+
|         1 | William | Turner | 1972-05-27 |
```

```
|         2 | Susan    | Smith  | 1975-11-02 |
+-----------+---------+--------+------------+
2 rows in set (0.00 sec)
```

Updating Data

When the data for William Turner was initially added to the table, data for the various address columns was omitted from the insert statement. The next statement shows how these columns can be populated via an update statement:

```
mysql> UPDATE person
    -> SET address = '1225 Tremont St.',
    ->   city = 'Boston',
    ->   state = 'MA',
    ->   country = 'USA',
    ->   postal_code = '02138'
    -> WHERE person_id = 1;
Query OK, 1 row affected (0.04 sec)
Rows matched: 1  Changed: 1  Warnings: 0
```

The server responded with a two-line message: the "Rows matched: 1" item tells you that the conditions of the where clause matched a single row in the table, and the "Changed: 1" item tells you that a single row in the table has been modified. Since the where clause specifies the primary key of William's row, this is exactly what you would expect to have happen.

As you can see, it is possible to modify more than one column in a single update statement. It is also possible to modify more than one row in a single statement depending on the conditions in your where clause. Consider, for example, what would happen if your where clause looked as follows:

```
WHERE person_id < 10
```

Since both William and Susan have a person_id value less than 10, both of their rows would be modified. If you leave off the where clause altogether, your update statement will modify every row in the table.

Deleting Data

It seems that William and Susan aren't getting along very well together, so one of them has got to go. Since William was there first, Susan will get the boot courtesy of the delete statement:

```
mysql> DELETE FROM person
    -> WHERE person_id = 2;
Query OK, 1 row affected (0.01 sec)
```

Again, the primary key is being used to isolate the row of interest, so a single row is deleted from the table. Similar to the update statement, more than one row can be deleted depending on the conditions in your where clause, and all rows will be deleted if the where clause is omitted.

When Good Statements Go Bad

So far, all of the SQL data statements shown in this chapter have been well-formed and have played by the rules. Based on the table definitions for the person and favorite_food tables, however, there are lots of ways that you can run afoul when inserting or modifying data. This section shows you some of the common mistakes that you might come across and how the MySQL server will respond.

Nonunique Primary Key

Because the table definitions include the creation of primary-key constraints, MySQL will make sure that duplicate values are not inserted into the tables. The next statement attempts to bypass the auto-increment feature of the person_id column and create another row in the person table with a person_id of 1:

```
mysql> INSERT INTO person
    -> (person_id, fname, lname, gender, birth_date)
    -> VALUES (1, 'Charles','Fulton', 'M', '1968-01-15');
ERROR 1062 (23000): Duplicate entry '1' for key 1
```

There is nothing stopping you (with the current schema objects, at least) from creating two rows with identical names, addresses, birth dates, etc., as long as they have different values for the person_id column.

Nonexistent Foreign Key

The table definition for the favorite_food table includes the creation of a foreign-key constraint on the person_id column. This constraint ensures that all values of person_id entered into the favorite_food table already exist in the person table. Here's what would happen if you tried to create a row that violates this constraint:

```
mysql> INSERT INTO favorite_food (person_id, food)
    -> VALUES (999, 'lasagna');
ERROR 1216 (23000): Cannot add or update a child row:
  a foreign key constraint fails
```

In this case, the favorite_food table is considered the *child*, and the person table is considered the *parent*, since the favorite_food table is dependent on the person table for its data. If you plan to enter data into both tables, you will need to create a row in parent before you can enter data into favorite_food.

> Foreign-key constraints are only enforced if your tables are created using the InnoDB storage engine. MySQL's storage engines will be discussed in Chapter 12.

Column Value Violations

The gender column in the person table is restricted to the values `'M'` for male and `'F'` for female. If you mistakenly attempt to set the value of the column to any other value, you will receive the following response:

```
mysql> UPDATE person
    -> SET gender = 'Z'
    -> WHERE person_id = 1;
Query OK, 1 row affected, 1 warning (0.01 sec)
Rows matched: 1  Changed: 1  Warnings: 1
```

The update statement did not fail, but it did produce a warning. To see the warning description, you can execute the show warnings command:

```
mysql> SHOW WARNINGS;
+---------+------+-------------------------------------------------+
| Level   | Code | Message                                         |
+---------+------+-------------------------------------------------+
| Warning | 1265 | Data truncated for column 'gender' at row 1     |
+---------+------+-------------------------------------------------+
1 row in set (0.00 sec)
```

This is what I would call a *soft error*, since the MySQL server didn't reject your statement but also didn't produce the expected results. In order to resolve the problem, the MySQL server sets the gender column to the empty string (''), which is definitely not what you intended. Personally, I would have preferred to have had the update statement rejected with an error message, which is what most other database servers would have done.

Invalid Date Conversions

If you construct a string with which to populate a date column, and that string does not match the expected format, you will receive another soft error. Here's an example that uses a date format that does not match the default date format of "YYYY-MM-DD":

```
mysql> UPDATE person
    -> SET birth_date = 'DEC-21-1980'
    -> WHERE person_id = 1;
Query OK, 1 rows affected, 1 warning (0.00 sec)
Rows matched: 1  Changed: 1  Warnings: 1
```

The show warnings command yields the following:

```
mysql> SHOW WARNINGS;
+---------+------+---------------------------------------------------+
| Level   | Code | Message                                           |
+---------+------+---------------------------------------------------+
| Warning | 1265 | Data truncated for column 'birth_date' at row 1   |
+---------+------+---------------------------------------------------+
```

Because it is a date column, `birth_date` cannot be set to an empty string, so MySQL sets the date to `'0000-00-00'`, as shown below:

```
mysql> SELECT birth_date
    -> FROM person
    -> WHERE person_id = 1;
+------------+
| birth_date |
+------------+
| 0000-00-00 |
+------------+
```

Again, I would prefer an error over a warning, since I now have invalid data (0000-00-00) in the person table.

The Bank Schema

For the remainder of the book, you will be using a group of tables that model a community bank. Some of the tables include Employee, Branch, Account, Customer, Product, Transaction, and Loan. The entire schema and example data should have been created when you followed the 13 steps at the beginning of the chapter for loading the MySQL server and generating the sample data. To see a diagram of the tables and their columns and relationships, see Appendix A.

Table 2-9 shows all of the tables used in the bank schema along with short definitions.

Table 2-9. Bank schema definitions

Table name	Definition
Account	A particular product opened for a particular customer
Business	A corporate customer (subtype of the Customer table)
Customer	A person or corporation known to the bank
Department	A group of bank employees implementing a particular banking function
Employee	A person working for the bank
Individual	A noncorporate customer (subtype of the Customer table)
Officer	A person allowed to transact business for a corporate customer
Product	A banking function offered to customers
Product_type	A group of products having similar function
Transaction	A change made to an account balance

Feel free to experiment with the tables as much as you want, including adding your own tables to expand the bank's business function. You can always drop the database and recreate it from the downloaded file if you want to make sure your sample data is intact.

If you want to see the tables available in your database, you can use the show tables command, as in:

```
mysql> SHOW TABLES;
+----------------+
| Tables_in_bank |
+----------------+
| account        |
| branch         |
| business       |
| customer       |
| department     |
| employee       |
| favorite_food  |
| individual     |
| officer        |
| person         |
| product        |
| product_type   |
| transaction    |
+----------------+
13 rows in set (0.10 sec)
```

Along with the 11 tables in the bank schema, the table listing also includes the two tables created in this chapter: person and favorite_food. These tables will not be used in later chapters, so feel free to drop them by issuing the following commands:

```
mysql> DROP TABLE favorite_food;
Query OK, 0 rows affected (0.56 sec)

mysql> DROP TABLE person;
Query OK, 0 rows affected (0.05 sec)
```

If you want to look at the columns in a table, you can use the describe command. Here's an example of the describe output for the customer table:

```
mysql> DESC customer;
+--------------+------------------+------+-----+---------+----------------+
| Field        | Type             | Null | Key | Default | Extra          |
+--------------+------------------+------+-----+---------+----------------+
| cust_id      | int(10) unsigned |      | PRI | NULL    | auto_increment |
| fed_id       | varchar(12)      |      |     |         |                |
| cust_type_cd | enum('I','B')    |      |     | I       |                |
| address      | varchar(30)      | YES  |     | NULL    |                |
| city         | varchar(20)      | YES  |     | NULL    |                |
| state        | varchar(20)      | YES  |     | NULL    |                |
| postal_code  | varchar(10)      | YES  |     | NULL    |                |
+--------------+------------------+------+-----+---------+----------------+
7 rows in set (0.03 sec)
```

The more comfortable you are with the example database, the better able you will be to understand the examples and, consequently, the concepts in the following chapters.

CHAPTER 3

Query Primer

So far, you have seen a few examples of database queries (a.k.a. select statements) sprinkled throughout the first two chapters. Now it's time to take a closer look at the different parts of the select statement and how they interact.

Query Mechanics

Before dissecting the select statement, it might be interesting to look at how queries are executed by the MySQL server (or, for that matter, any database server). If you are using the *mysql* command-line tool (which I assume you are), then you have already logged in to the MySQL server by providing your username and password (and possibly a hostname if the MySQL server is running on a different computer). Once the server has verified that your username and password are correct, a *database connection* is generated for you to use. This connection is held by the application that requested it (which, in this case, is the *mysql* tool) until either the application releases the connection (i.e., as a result of your typing *quit*) or the server closes the connection (i.e., when the server is shut down). Each connection to the MySQL server is assigned an identifier, which is shown to you when you first log in:

```
Welcome to the MySQL monitor. Commands end with ; or \g.
Your MySQL connection id is 2 to server version: 4.1.11-nt

Type 'help;' or '\h' for help. Type '\c' to clear the buffer.
```

In this case, my connection ID is 2. This information might be useful to your database administrator if something goes awry, such as a malformed query that runs for hours, so you might want to jot it down.

Once the server has verified your username and password and issued you a connection, you are ready to execute queries (along with other SQL statements). Each time a query is sent to the server, the server checks the following things prior to statement execution:

- Do you have permission to execute the statement?
- Do you have permission to access the desired data?
- Is your statement syntax correct?

If your statement passes these three tests, then your query is handed to the *query optimizer*, whose job it is to determine the most efficient way to execute your query. The optimizer will look at such things as the order in which to join the tables named in the query and what indexes are available, and then picks an *execution plan*, which is used by the server to execute your query.

 Understanding and influencing how your database server chooses execution plans is a fascinating topic that many of you will wish to explore. For those readers using MySQL, you might consider reading *High Performance MySQL* (O'Reilly). Among other things, you will learn how to generate indexes, analyze execution plans, influence the optimizer via query hints, and tune your server's startup parameters. If you are using Oracle Database or SQL Server, there are dozens of tuning books available.

Once the server has finished executing your query, the *result set* is returned to the calling application (which is, once again, the *mysql* tool). As was mentioned in Chapter 1, a result set is just another table containing rows and columns. If your query fails to yield any results, the *mysql* tool will show you the message found at the end of the following example:

```
mysql> SELECT emp_id, fname, lname
    -> FROM employee
    -> WHERE lname = 'Bkadfl';
Empty set (0.00 sec)
```

If the query returns one or more rows, the *mysql* tool will format the results by adding column headers and by constructing boxes around the columns using the -, |, and + symbols, as shown in the next example:

```
mysql> SELECT fname, lname
    -> FROM employee;
+----------+-----------+
| fname    | lname     |
+----------+-----------+
| Michael  | Smith     |
| Susan    | Barker    |
| Robert   | Tyler     |
| Susan    | Hawthorne |
| John     | Gooding   |
| Helen    | Fleming   |
| Chris    | Tucker    |
| Sarah    | Parker    |
```

```
| Jane     | Grossman  |
| Paula    | Roberts   |
| Thomas   | Ziegler   |
| Samantha | Jameson   |
| John     | Blake     |
| Cindy    | Mason     |
| Frank    | Portman   |
| Theresa  | Markham   |
| Beth     | Fowler    |
| Rick     | Tulman    |
+----------+-----------+
18 rows in set (0.00 sec)
```

This query returns the first and last names of all of the employees in the employee table. After the last row of data is displayed, the *mysql* tool displays a message telling you how many rows were returned, which, in this case, is 18.

Query Clauses

There are several components or *clauses* that make up the select statement. While only one of them is mandatory when using MySQL (the select clause), you will usually include at least two or three of the six available clauses. Table 3-1 shows the different clauses and their purposes.

Table 3-1. Query clauses

Clause name	Purpose
Select	Determines which columns to include in the query's result set
From	Identifies the tables from which to draw data and how the tables should be joined
Where	Restricts the number of rows in the final result set
Group by	Used to group rows together by common column values
Having	Restricts the number of rows in the final result set using grouped data
Order by	Sorts the rows of the final result set by one or more columns

All of the clauses shown in Table 3-1 are included in the ANSI specification; additionally, there are several other clauses unique to MySQL that will be explored in Appendix B. The following sections delve into the uses of the six major query clauses.

The select Clause

Even though the select clause is the first clause of a select statement, it is one of the last clauses to be evaluated by the database server. The reason for this is that before you can determine what to include in the final result set, you need to know all of the possible columns that *could* be included in the final result set. In order to fully understand the role of the select clause, therefore, you will need to understand a bit about the from clause. Here's a query to get started:

```
mysql> SELECT *
    -> FROM department;
+---------+----------------+
| dept_id | name           |
+---------+----------------+
|       1 | Operations     |
|       2 | Loans          |
|       3 | Administration |
+---------+----------------+
3 rows in set (0.04 sec)
```

In this query, the from clause lists a single table (department), and the select clause indicates that *all* columns (designated by "*") in the department table should be included in the result set. This query could be described in English as follows:

Show me all the columns in the department table.

In addition to specifying all of the columns via the asterisk character, you can explicitly name the columns you are interested in, such as:

```
mysql> SELECT dept_id, name
    -> FROM department;
+---------+----------------+
| dept_id | name           |
+---------+----------------+
|       1 | Operations     |
|       2 | Loans          |
|       3 | Administration |
+---------+----------------+
3 rows in set (0.01 sec)
```

The results are identical to the first query, since all of the columns in the department table (dept_id and name) are named in the select clause. You can choose to include only a subset of the columns in the department table as well:

```
mysql> SELECT name
    -> FROM department;
+----------------+
| name           |
+----------------+
| Operations     |
| Loans          |
| Administration |
+----------------+
3 rows in set (0.00 sec)
```

The job of the select clause, therefore, is the following:

> The select clause determines which of all possible columns should be included in the query's result set.

If you were limited to including only columns from the table or tables named in the from clause, things would be rather dull. However, you can spice things up by including in your select clause such things as:

- Literals, such as numbers or strings
- Expressions, such as transaction.amount * -1
- Built-in function calls, such as ROUND(transaction.amount, 2)

The next query demonstrates the use of a table column, a literal, an expression, and a built-in function call in a single query against the employee table:

```
mysql> SELECT emp_id,
    -> 'ACTIVE',
    -> emp_id * 3.14159,
    -> UPPER(lname)
    -> FROM employee;
+--------+--------+------------------+--------------+
| emp_id | ACTIVE | emp_id * 3.14159 | UPPER(lname) |
+--------+--------+------------------+--------------+
|      1 | ACTIVE |          3.14159 | SMITH        |
|      2 | ACTIVE |          6.28318 | BARKER       |
|      3 | ACTIVE |          9.42477 | TYLER        |
|      4 | ACTIVE |         12.56636 | HAWTHORNE    |
|      5 | ACTIVE |         15.70795 | GOODING      |
|      6 | ACTIVE |         18.84954 | FLEMING      |
|      7 | ACTIVE |         21.99113 | TUCKER       |
|      8 | ACTIVE |         25.13272 | PARKER       |
|      9 | ACTIVE |         28.27431 | GROSSMAN     |
|     10 | ACTIVE |         31.41590 | ROBERTS      |
|     11 | ACTIVE |         34.55749 | ZIEGLER      |
|     12 | ACTIVE |         37.69908 | JAMESON      |
|     13 | ACTIVE |         40.84067 | BLAKE        |
|     14 | ACTIVE |         43.98226 | MASON        |
|     15 | ACTIVE |         47.12385 | PORTMAN      |
|     16 | ACTIVE |         50.26544 | MARKHAM      |
|     17 | ACTIVE |         53.40703 | FOWLER       |
|     18 | ACTIVE |         56.54862 | TULMAN       |
+--------+--------+------------------+--------------+
18 rows in set (0.05 sec)
```

Expressions and built-in functions will be covered in detail later, but I wanted to give you a feel for what kinds of things can be included in the select clause. If you only need to execute a built-in function or evaluate a simple expression, you can skip the from clause entirely. Here's an example:

```
mysql> SELECT VERSION(),
    ->     USER(),
    ->     DATABASE();
```

```
+-----------+-------------------+------------+
| VERSION() | USER()            | DATABASE() |
+-----------+-------------------+------------+
| 4.1.11-nt | lrngsql@localhost | bank       |
+-----------+-------------------+------------+
1 row in set (0.02 sec)
```

Since this query simply calls three built-in functions and doesn't retrieve data from any tables, there is no need for a from clause.

Column Aliases

Although the *mysql* tool will generate labels for the columns returned by your queries, you may want to assign your own labels. While you might want to assign a new label to a column from a table (if it is poorly or ambiguously named), you will almost certainly want to assign your own labels to those columns in your result set that are generated by expressions or built-in function calls. You can do so by adding a *column alias* after each element of your select clause. Here's the previous query against the employee table with column aliases applied to three of the columns:

```
mysql> SELECT emp_id,
    ->   'ACTIVE' status,
    ->   emp_id * 3.14159 empid_x_pi,
    ->   UPPER(lname) last_name_upper
    -> FROM employee;
+--------+--------+------------+-----------------+
| emp_id | status | empid_x_pi | last_name_upper |
+--------+--------+------------+-----------------+
|      1 | ACTIVE |    3.14159 | SMITH           |
|      2 | ACTIVE |    6.28318 | BARKER          |
|      3 | ACTIVE |    9.42477 | TYLER           |
|      4 | ACTIVE |   12.56636 | HAWTHORNE       |
|      5 | ACTIVE |   15.70795 | GOODING         |
|      6 | ACTIVE |   18.84954 | FLEMING         |
|      7 | ACTIVE |   21.99113 | TUCKER          |
|      8 | ACTIVE |   25.13272 | PARKER          |
|      9 | ACTIVE |   28.27431 | GROSSMAN        |
|     10 | ACTIVE |   31.41590 | ROBERTS         |
|     11 | ACTIVE |   34.55749 | ZIEGLER         |
|     12 | ACTIVE |   37.69908 | JAMESON         |
|     13 | ACTIVE |   40.84067 | BLAKE           |
|     14 | ACTIVE |   43.98226 | MASON           |
|     15 | ACTIVE |   47.12385 | PORTMAN         |
|     16 | ACTIVE |   50.26544 | MARKHAM         |
|     17 | ACTIVE |   53.40703 | FOWLER          |
|     18 | ACTIVE |   56.54862 | TULMAN          |
+--------+--------+------------+-----------------+
18 rows in set (0.00 sec)
```

If you look at the column headers, you can see that the second, third, and fourth columns now have reasonable names instead of simply being labeled with the function or expression that generated the column. If you look at the select clause, you can see how the column aliases `status`, `empid_x_pi`, and `last_name_upper` are added after the second, third, and fourth columns. I think you will agree that the output is easier to understand with column aliases in place, and it would be easier to work with programmatically if you were issuing the query from within Java or C# rather than interactively via the *mysql* tool.

Removing Duplicates

In some cases, a query might return duplicate rows of data. For example, if you were to retrieve the ID's of all customers that have accounts, you would see the following:

```
mysql> SELECT cust_id
    -> FROM account;
+---------+
| cust_id |
+---------+
|       1 |
|       1 |
|       1 |
|       2 |
|       2 |
|       3 |
|       3 |
|       4 |
|       4 |
|       4 |
|       5 |
|       6 |
|       6 |
|       7 |
|       8 |
|       8 |
|       9 |
|       9 |
|       9 |
|      10 |
|      10 |
|      11 |
|      12 |
|      13 |
+---------+
24 rows in set (0.00 sec)
```

Since some customers have more than one account, you will see the same customer ID once for each account owned by that customer. What you probably want in this case is the *distinct* set of customers that have accounts, instead of seeing the customer ID for each row in the account table. You can achieve this by adding the keyword `distinct` directly after the `select` keyword, as the following shows:

```
mysql> SELECT DISTINCT cust_id
    -> FROM account;
+---------+
```

```
| cust_id |
+---------+
|       1 |
|       2 |
|       3 |
|       4 |
|       5 |
|       6 |
|       7 |
|       8 |
|       9 |
|      10 |
|      11 |
|      12 |
|      13 |
+---------+
13 rows in set (0.01 sec)
```

The result set now contains 13 rows, one for each distinct customer, rather than 24 rows, one for each account.

If you do not want the server to remove duplicate data, or you are sure there will be no duplicates in your result set, you can specify the ALL keyword instead of specifying DISTINCT. However, the ALL keyword is the default and never needs to be explicitly named, so most programmers do not include ALL in their queries.

 Remember that generating a distinct set of results requires the data to be sorted, which can be time consuming for large result sets. Don't fall into the trap of using DISTINCT just to be sure there are no duplicates; instead, take the time to understand the data you are working with so you will know whether duplicates are possible.

The from Clause

Thus far, you have seen queries whose from clauses contain a single table. Although most SQL books will define the from clause as simply a list of one or more tables, I would like to broaden the definition as follows:

> The from clause defines the tables used by a query, along with the means of linking the tables together.

This definition is composed of two separate but related concepts, which will be explored in the following sections.

Tables

When confronted with the term table, most people think of a set of related rows stored in a database. While this does describe one type of table, I would like to use the word in a more general way by removing any notion of how the data might be stored and concentrating on just the set of related rows. There are three different types of tables that meet this relaxed definition:

- Permanent tables (i.e., created using the create table statement)
- Temporary tables (i.e., rows returned by a subquery)
- Virtual tables (i.e., created using the create view statement)

Each of these table types may be included in a query's from clause. By now, you should be comfortable with including a permanent table in a from clause, so I will briefly describe the other types of tables that can be referenced in a from clause.

Subquery-generated tables

A subquery is a query contained within another query. Subqueries are surrounded by parentheses and can be found in various parts of a select statement; within the from clause, however, a subquery serves the role of generating a temporary table that is visible from all other query clauses and can interact with other tables named in the from clause. Here's a simple example:

```
mysql> SELECT e.emp_id, e.fname, e.lname
    -> FROM (SELECT emp_id, fname, lname, start_date, title
    ->   FROM employee) e;
+--------+----------+-----------+
| emp_id | fname    | lname     |
+--------+----------+-----------+
|      1 | Michael  | Smith     |
|      2 | Susan    | Barker    |
|      3 | Robert   | Tyler     |
|      4 | Susan    | Hawthorne |
|      5 | John     | Gooding   |
|      6 | Helen    | Fleming   |
|      7 | Chris    | Tucker    |
|      8 | Sarah    | Parker    |
|      9 | Jane     | Grossman  |
|     10 | Paula    | Roberts   |
|     11 | Thomas   | Ziegler   |
|     12 | Samantha | Jameson   |
|     13 | John     | Blake     |
|     14 | Cindy    | Mason     |
|     15 | Frank    | Portman   |
|     16 | Theresa  | Markham   |
|     17 | Beth     | Fowler    |
|     18 | Rick     | Tulman    |
+--------+----------+-----------+
18 rows in set (0.00 sec)
```

In this example, a subquery against the employee table returns five columns, and the *containing query* references three of the five available columns. The subquery is referenced by the containing query via its alias, which, in this case, is e. This is a simplistic and not-particularly-useful example of a subquery in a from clause; you will find complete coverage of subqueries in Chapter 9.

Views

A view is a query that is stored in the data dictionary. It looks and acts like a table, but there is no data associated with a view (this is why I call it a *virtual* table). When you issue a query against a view, your query is merged with the view definition to create a final query to be executed.

To demonstrate, here's a view definition that queries the employee table and includes a call to a built-in function:

```
CREATE VIEW employee_vw AS
SELECT emp_id, fname, lname,
  YEAR(start_date) start_year
FROM employee;
```

After the view has been created, no additional data is created: the select statement is simply stored by the server for future use. Now that the view exists, you can issue queries against it, as in:

```
SELECT emp_id, start_year
FROM employee_vw;
```

```
emp_id  start_year
-------- ----------
       1       2001
       2       2002
       3       2000
       4       2002
       5       2003
       6       2004
       7       2004
       8       2002
       9       2002
      10       2002
      11       2000
      12       2003
      13       2000
      14       2002
      15       2003
      16       2001
      17       2002
      18       2002
```

Views are created for various reasons, including to hide columns from users and to simplify complex database designs.

Views are not supported in MySQL until Version 5.0.1. They are, however, a heavily used feature of the other database servers, so, for those readers planning to work with MySQL, you should keep them in mind.

Because Version 4.1.11 of MySQL does not include views, I intentionally left out the *mysql>* prompt in the previous query along with the usual result-set formatting. I use this same convention in several other chapters in the book when I show a SQL feature not yet implemented by MySQL.

Table Links

The second deviation from the simple from clause definition is the mandate that if more than one table appears in the from clause, then the conditions used to *link* the tables must be included as well. This is not a requirement of MySQL or any other database server, but it is the ANSI-approved method of joining multiple tables, and it is the most portable across the various database servers. Joining multiple tables will be explored in depth in Chapters 5 and 10, but here's a simple example in case I have piqued your curiosity:

```
mysql> SELECT employee.emp_id, employee.fname,
    ->    employee.lname, department.name dept_name
    -> FROM employee INNER JOIN department
    ->    ON employee.dept_id = department.dept_id;
+--------+----------+-----------+----------------+
| emp_id | fname    | lname     | dept_name      |
+--------+----------+-----------+----------------+
|      1 | Michael  | Smith     | Administration |
|      2 | Susan    | Barker    | Administration |
|      3 | Robert   | Tyler     | Administration |
|      4 | Susan    | Hawthorne | Operations     |
|      5 | John     | Gooding   | Loans          |
|      6 | Helen    | Fleming   | Operations     |
|      7 | Chris    | Tucker    | Operations     |
|      8 | Sarah    | Parker    | Operations     |
|      9 | Jane     | Grossman  | Operations     |
|     10 | Paula    | Roberts   | Operations     |
|     11 | Thomas   | Ziegler   | Operations     |
|     12 | Samantha | Jameson   | Operations     |
|     13 | John     | Blake     | Operations     |
|     14 | Cindy    | Mason     | Operations     |
|     15 | Frank    | Portman   | Operations     |
|     16 | Theresa  | Markham   | Operations     |
|     17 | Beth     | Fowler    | Operations     |
|     18 | Rick     | Tulman    | Operations     |
+--------+----------+-----------+----------------+
18 rows in set (0.05 sec)
```

The previous query displays data from both the employee table (emp_id, fname, lname) and the department table (name), so both tables are included in the from clause. The mechanism for linking the two tables (referred to as a *join*) is the employee's department affiliation stored in the employee table. Thus, the database server is instructed to use the value of the dept_id column in the employee table to look up the associated department name in the department table. Join conditions for two tables are found in the on subclause of the from clause; in this case, the join condition is ON e.dept_id = d.dept_id. Again, please refer to Chapter 5 for a thorough discussion of joining multiple tables.

Defining Table Aliases

When multiple tables are joined in a single query, you need a way to identify which table you are referring to when you reference columns in the select, where, group by, having, and order by clauses. You have two choices when referencing a table outside the from clause:

- Use the entire table name, such as employee.emp_id.
- Assign each table an *alias* and use the alias throughout the query.

In the previous query, I chose to use the entire table name in the select and on clauses. Here's what the same query looks like using table aliases:

```
SELECT e.emp_id, e.fname, e.lname,
  d.name dept_name
FROM employee e INNER JOIN department d
  ON e.dept_id = d.dept_id;
```

If you look closely at the from clause, you will see that the employee table is assigned the alias e, and the department table is assigned the alias d. These aliases are then used in the on clause when defining the join condition as well as in the select clause when specifying the columns to include in the result set. I hope you will agree that using aliases makes for a more compact statement without causing confusion (as long as your choices for alias names are reasonable).

The where Clause

The queries shown thus far in the chapter have selected every row from the employee, department, or account tables (except for the demonstration of distinct earlier in the chapter). Most of the time, however, you will not wish to retrieve *every* row from a table but will want a way to filter out those rows that are not of interest. This is a job for the where clause.

 The where clause is the mechanism for filtering out unwanted rows from your result set.

For example, perhaps you are interested in retrieving data from the employee table, but only for those employees who are employed as head tellers. The following query employs a where clause to retrieve *only* the four head tellers:

```
mysql> SELECT emp_id, fname, lname, start_date, title
    -> FROM employee
    -> WHERE title = 'Head Teller';
+--------+---------+---------+------------+-------------+
| emp_id | fname   | lname   | start_date | title       |
+--------+---------+---------+------------+-------------+
|      6 | Helen   | Fleming | 2004-03-17 | Head Teller |
|     10 | Paula   | Roberts | 2002-07-27 | Head Teller |
|     13 | John    | Blake   | 2000-05-11 | Head Teller |
|     16 | Theresa | Markham | 2001-03-15 | Head Teller |
+--------+---------+---------+------------+-------------+
4 rows in set (0.00 sec)
```

In this case, 14 of the 18 employee rows were filtered out by the where clause. This where clause contains a single *filter condition*, but you can include as many conditions as required; individual conditions are separated using operators such as and, or, and not (see Chapter 4 for a complete discussion of the where clause and filter conditions). Here's an extension of the previous query that includes a second condition stating that only those employees with a start date later than January 1, 2002, should be included:

```
mysql> SELECT emp_id, fname, lname, start_date, title
    -> FROM employee
    -> WHERE title = 'Head Teller'
    ->   AND start_date > '2002-01-01';
+--------+-------+---------+------------+-------------+
| emp_id | fname | lname   | start_date | title       |
+--------+-------+---------+------------+-------------+
|      6 | Helen | Fleming | 2004-03-17 | Head Teller |
|     10 | Paula | Roberts | 2002-07-27 | Head Teller |
+--------+-------+---------+------------+-------------+
2 rows in set (0.00 sec)
```

The first condition (title = 'Head Teller') filtered out 14 of 18 employee rows, and the second condition (start_date > '2002-01-01') filtered out an additional 2 rows, leaving 2 rows in the result set. Let's see what would happen if you change the operator separating the two conditions from and to or:

```
mysql> SELECT emp_id, fname, lname, start_date, title
    -> FROM employee
    -> WHERE title = 'Head Teller'
    ->   OR start_date > '2002-01-01';
+--------+---------+-----------+------------+--------------------+
| emp_id | fname   | lname     | start_date | title              |
+--------+---------+-----------+------------+--------------------+
|      2 | Susan   | Barker    | 2002-09-12 | Vice President     |
|      4 | Susan   | Hawthorne | 2002-04-24 | Operations Manager |
|      5 | John    | Gooding   | 2003-11-14 | Loan Manager       |
|      6 | Helen   | Fleming   | 2004-03-17 | Head Teller        |
```

```
|       7 | Chris    | Tucker    | 2004-09-15 | Teller      |
|       8 | Sarah    | Parker    | 2002-12-02 | Teller      |
|       9 | Jane     | Grossman  | 2002-05-03 | Teller      |
|      10 | Paula    | Roberts   | 2002-07-27 | Head Teller |
|      12 | Samantha | Jameson   | 2003-01-08 | Teller      |
|      13 | John     | Blake     | 2000-05-11 | Head Teller |
|      14 | Cindy    | Mason     | 2002-08-09 | Teller      |
|      15 | Frank    | Portman   | 2003-04-01 | Teller      |
|      16 | Theresa  | Markham   | 2001-03-15 | Head Teller |
|      17 | Beth     | Fowler    | 2002-06-29 | Teller      |
|      18 | Rick     | Tulman    | 2002-12-12 | Teller      |
+--------+----------+-----------+------------+--------------------+
15 rows in set (0.00 sec)
```

Looking at the output, you can see that all four Head Tellers are included in the result set, along with any other employee who started working for the bank after January 1, 2002. At least one of the two conditions is true for 15 of the 18 employees in the employee table. Thus, when you separate conditions using the and operator, *all* conditions must evaluate to true to be included in the result set; when you use or, however, only *one* of the conditions need evaluate to true for a row to be included.

So what should you do if you need to use both and and or operators in your where clause? Glad you asked. You should use parentheses to group conditions together. The next query specifies that only those employees who are Head Tellers and began working for the company after January 1, 2002, *or* those employees who are Tellers and began working after January 1, 2003, be included in the result set:

```
mysql> SELECT emp_id, fname, lname, start_date, title
    -> FROM employee
    -> WHERE (title = 'Head Teller' AND start_date > '2002-01-01')
    ->   OR (title = 'Teller' AND start_date > '2003-01-01');
+--------+----------+---------+------------+-------------+
| emp_id | fname    | lname   | start_date | title       |
+--------+----------+---------+------------+-------------+
|      6 | Helen    | Fleming | 2004-03-17 | Head Teller |
|      7 | Chris    | Tucker  | 2004-09-15 | Teller      |
|     10 | Paula    | Roberts | 2002-07-27 | Head Teller |
|     12 | Samantha | Jameson | 2003-01-08 | Teller      |
|     15 | Frank    | Portman | 2003-04-01 | Teller      |
+--------+----------+---------+------------+-------------+
5 rows in set (0.00 sec)
```

You should always use parentheses to separate groups of conditions when mixing different operators so that you, the database server, and anyone who comes along later to modify your code will be on the same page.

The group by and having Clauses

All of the queries thus far have retrieved raw data without any manipulation. Sometimes, however, you will want to find trends in your data that will require the database server to cook the data a bit before you retrieve your result set. One such mechanism is the group by clause, which is used to group data by column values. For example, rather than looking at a list of employees and the departments to which they are assigned, you might want to look at a list of departments along with the number of employees assigned to each department. When using the group by clause, you may also use the having clause, which allows you to filter group data in the same way the where clause lets you filter raw data.

I wanted to briefly mention these two clauses so they don't catch you by surprise later in the book, but they are a bit more advanced than the other four select clauses. Therefore, I ask that you wait until Chapter 8 for a full description of how and when to use group by and having.

The order by Clause

In general, the rows in a result set returned from a query are not in any particular order. If you want to your result set in a particular order, you will need to instruct the server to sort the results using the order by clause:

> The order by clause is the mechanism for sorting your result set using either raw column data or expressions based on column data.

For example, here's another look at an earlier query against the account table:

```
mysql> SELECT open_emp_id, product_cd
    -> FROM account;
+-------------+------------+
| open_emp_id | product_cd |
+-------------+------------+
|          10 | CHK        |
|          10 | SAV        |
|          10 | CD         |
|          10 | CHK        |
|          10 | SAV        |
|          13 | CHK        |
|          13 | MM         |
|           1 | CHK        |
|           1 | SAV        |
|           1 | MM         |
|          16 | CHK        |
|           1 | CHK        |
|           1 | CD         |
|          10 | CD         |
|          16 | CHK        |
|          16 | SAV        |
|           1 | CHK        |
|           1 | MM         |
|           1 | CD         |
|          16 | CHK        |
|          16 | BUS        |
|          10 | BUS        |
```

```
|          16 | CHK        |
|          13 | SBL        |
+-------------+------------+
24 rows in set (0.00 sec)
```

If you are trying to analyze data for each employee, it would be helpful to sort the results by the open_emp_id column; to do so, simply add this column to the order by clause:

```
mysql> SELECT open_emp_id, product_cd
    -> FROM account
    -> ORDER BY open_emp_id;
+-------------+------------+
| open_emp_id | product_cd |
+-------------+------------+
|           1 | CHK        |
|           1 | SAV        |
|           1 | MM         |
|           1 | CHK        |
|           1 | CD         |
|           1 | CHK        |
|           1 | MM         |
|           1 | CD         |
|          10 | CHK        |
|          10 | SAV        |
|          10 | CD         |
|          10 | CHK        |
|          10 | SAV        |
|          10 | CD         |
|          10 | BUS        |
|          13 | CHK        |
|          13 | MM         |
|          13 | SBL        |
|          16 | CHK        |
|          16 | CHK        |
|          16 | SAV        |
|          16 | CHK        |
|          16 | BUS        |
|          16 | CHK        |
+-------------+------------+
24 rows in set (0.00 sec)
```

It is now easier to see what types of accounts were opened by each employee. However, it might be even better if you could ensure that the account types were shown in the same order for each distinct employee; this can be accomplished by adding the product_cd column after the open_emp_id column in the order by clause:

```
mysql> SELECT open_emp_id, product_cd
    -> FROM account
    -> ORDER BY open_emp_id, product_cd;
+-------------+------------+
| open_emp_id | product_cd |
+-------------+------------+
|           1 | CD         |
```

```
|        1 | CD          |
|        1 | CHK         |
|        1 | CHK         |
|        1 | CHK         |
|        1 | MM          |
|        1 | MM          |
|        1 | SAV         |
|       10 | BUS         |
|       10 | CD          |
|       10 | CD          |
|       10 | CHK         |
|       10 | CHK         |
|       10 | SAV         |
|       10 | SAV         |
|       13 | CHK         |
|       13 | MM          |
|       13 | SBL         |
|       16 | BUS         |
|       16 | CHK         |
|       16 | CHK         |
|       16 | CHK         |
|       16 | CHK         |
|       16 | SAV         |
+------------+-----------+
24 rows in set (0.00 sec)
```

The result set has now been sorted first by employee ID and then secondly by the account type. The order that columns appear in your order by clause does make a difference.

Ascending Versus Descending Sort Order

When sorting, you have the option of specifying *ascending* or *descending* order via the asc and desc keywords. The default is ascending, so you will only need to add the desc keyword if you want to use a descending sort. For example, the following query lists all accounts sorted by available balance with the highest balance listed at the top:

```
mysql> SELECT account_id, product_cd, open_date, avail_balance
    -> FROM account
    -> ORDER BY avail_balance DESC;
+------------+------------+------------+---------------+
| account_id | product_cd | open_date  | avail_balance |
+------------+------------+------------+---------------+
|         24 | SBL        | 2004-02-22 |      50000.00 |
|         23 | CHK        | 2003-07-30 |      38552.05 |
|         20 | CHK        | 2002-09-30 |      23575.12 |
|         13 | CD         | 2004-12-28 |      10000.00 |
|         22 | BUS        | 2004-03-22 |       9345.55 |
|         18 | MM         | 2004-10-28 |       9345.55 |
|         10 | MM         | 2004-09-30 |       5487.09 |
|         14 | CD         | 2004-01-12 |       5000.00 |
|         15 | CHK        | 2001-05-23 |       3487.19 |
```

```
|           3 | CD          | 2004-06-30 |        3000.00 |
|           4 | CHK         | 2001-03-12 |        2258.02 |
|          11 | CHK         | 2004-01-27 |        2237.97 |
|           7 | MM          | 2002-12-15 |        2212.50 |
|          19 | CD          | 2004-06-30 |        1500.00 |
|           1 | CHK         | 2000-01-15 |        1057.75 |
|           6 | CHK         | 2002-11-23 |        1057.75 |
|           9 | SAV         | 2000-01-15 |         767.77 |
|           8 | CHK         | 2003-09-12 |         534.12 |
|           2 | SAV         | 2000-01-15 |         500.00 |
|          16 | SAV         | 2001-05-23 |         387.99 |
|           5 | SAV         | 2001-03-12 |         200.00 |
|          17 | CHK         | 2003-07-30 |         125.67 |
|          12 | CHK         | 2002-08-24 |         122.37 |
|          21 | BUS         | 2002-10-01 |           0.00 |
+-------------+-------------+------------+----------------+
24 rows in set (0.01 sec)
```

Descending sorts are commonly used for ranking queries, such as "show me the top 5 account balances." MySQL includes a limit clause that allows you to sort your data and then discard all but the first *X* rows; see Appendix B for a discussion of the limit clause, along with other non-ANSI extensions.

Sorting via Expressions

Sorting your results using column data is all well and good, but sometimes you might need to sort by something that is not stored in the database, and possibly doesn't appear anywhere in your query. You can add an expression to your order by clause to handle such situations. For example, perhaps you would like to sort your customer data by the last three digits of the customer's Federal ID number (which is either a Social Security number for individuals or a corporate ID for businesses):

```
mysql> SELECT cust_id, cust_type_cd, city, state, fed_id
    -> FROM customer
    -> ORDER BY RIGHT(fed_id, 3);
+---------+--------------+------------+-------+-------------+
| cust_id | cust_type_cd | city       | state | fed_id      |
+---------+--------------+------------+-------+-------------+
|       1 | I            | Lynnfield  | MA    | 111-11-1111 |
|      10 | B            | Salem      | NH    | 04-1111111  |
|       2 | I            | Woburn     | MA    | 222-22-2222 |
|      11 | B            | Wilmington | MA    | 04-2222222  |
|       3 | I            | Quincy     | MA    | 333-33-3333 |
|      12 | B            | Salem      | NH    | 04-3333333  |
|      13 | B            | Quincy     | MA    | 04-4444444  |
|       4 | I            | Waltham    | MA    | 444-44-4444 |
|       5 | I            | Salem      | NH    | 555-55-5555 |
|       6 | I            | Waltham    | MA    | 666-66-6666 |
|       7 | I            | Wilmington | MA    | 777-77-7777 |
|       8 | I            | Salem      | NH    | 888-88-8888 |
|       9 | I            | Newton     | MA    | 999-99-9999 |
+---------+--------------+------------+-------+-------------+
13 rows in set (0.01 sec)
```

This query uses the built-in function `right( )` to extract the last three characters of the fed_id column and then sorts the rows based on this value.

Sorting via Numeric Placeholders

If you are sorting using the columns in your select clause, you can opt to reference the columns by their *position* in the select clause rather than by name. For example, if you want to sort using the second and fifth columns returned by a query, you could do the following:

```
mysql> SELECT emp_id, title, start_date, fname, lname
    -> FROM employee
    -> ORDER BY 2, 5;
+--------+-------------------+------------+----------+----------+
| emp_id | title             | start_date | fname    | lname    |
+--------+-------------------+------------+----------+----------+
|     13 | Head Teller       | 2000-05-11 | John     | Blake    |
|      6 | Head Teller       | 2004-03-17 | Helen    | Fleming  |
|     16 | Head Teller       | 2001-03-15 | Theresa  | Markham  |
|     10 | Head Teller       | 2002-07-27 | Paula    | Roberts  |
|      5 | Loan Manager      | 2003-11-14 | John     | Gooding  |
|      4 | Operations Manager| 2002-04-24 | Susan    | Hawthorne|
|      1 | President         | 2001-06-22 | Michael  | Smith    |
|     17 | Teller            | 2002-06-29 | Beth     | Fowler   |
|      9 | Teller            | 2002-05-03 | Jane     | Grossman |
|     12 | Teller            | 2003-01-08 | Samantha | Jameson  |
|     14 | Teller            | 2002-08-09 | Cindy    | Mason    |
|      8 | Teller            | 2002-12-02 | Sarah    | Parker   |
|     15 | Teller            | 2003-04-01 | Frank    | Portman  |
|      7 | Teller            | 2004-09-15 | Chris    | Tucker   |
|     18 | Teller            | 2002-12-12 | Rick     | Tulman   |
|     11 | Teller            | 2000-10-23 | Thomas   | Ziegler  |
|      3 | Treasurer         | 2000-02-09 | Robert   | Tyler    |
|      2 | Vice President    | 2002-09-12 | Susan    | Barker   |
+--------+-------------------+------------+----------+----------+
18 rows in set (0.03 sec)
```

You might want to use this feature sparingly, since adding a column to the select clause without changing the numbers in the order by clause can lead to unexpected results.

Exercises

The following exercises are designed to strengthen your understanding of the select statement and its various clauses. Please see Appendix C for solutions.

3-1

Retrieve the employee ID, first name, and last name for all bank employees. Sort by last name then first name.

3-2

Retrieve the account ID, customer ID, and available balance for all accounts whose status equals 'ACTIVE' and whose available balance is greater than $2,500.

3-3

Write a query against the account table that returns the IDs of the employees who opened the accounts (use the account.open_emp_id column). Include a single row for each distinct employee.

3-4

Fill in the blanks (denoted by <#>) for this multi-data-set query to achieve the results shown below:

```
mysql> SELECT p.product_cd, a.cust_id, a.avail_balance
    -> FROM product p INNER JOIN account <1>
    ->    ON p.product_cd = <2>
    -> WHERE p.<3> = 'ACCOUNT';
+------------+---------+---------------+
| product_cd | cust_id | avail_balance |
+------------+---------+---------------+
| CD         |       1 |       3000.00 |
| CD         |       6 |      10000.00 |
| CD         |       7 |       5000.00 |
| CD         |       9 |       1500.00 |
| CHK        |       1 |       1057.75 |
| CHK        |       2 |       2258.02 |
| CHK        |       3 |       1057.75 |
| CHK        |       4 |        534.12 |
| CHK        |       5 |       2237.97 |
| CHK        |       6 |        122.37 |
| CHK        |       8 |       3487.19 |
| CHK        |       9 |        125.67 |
| CHK        |      10 |      23575.12 |
| CHK        |      12 |      38552.05 |
| MM         |       3 |       2212.50 |
| MM         |       4 |       5487.09 |
| MM         |       9 |       9345.55 |
| SAV        |       1 |        500.00 |
| SAV        |       2 |        200.00 |
| SAV        |       4 |        767.77 |
| SAV        |       8 |        387.99 |
+------------+---------+---------------+
21 rows in set (0.02 sec)
```

CHAPTER 4
Filtering

There are some instances when you will want to work with every row in a table, such as:

- Purging all data from a table used to stage new data warehouse feeds.
- Modifying all rows in a table after a new column has been added.
- Retrieving all rows from a message queue table.

In cases like these, your SQL statements won't need to have a where clause, since you don't need to exclude any rows from consideration. Most of the time, however, you will want to narrow your focus to a subset of a table's rows. Therefore, all of the SQL data statements (except the insert statement) include an optional where clause to house all filter conditions used to restrict the number of rows acted on by the SQL statement. Additionally, the select statement includes a having clause in which filter conditions pertaining to grouped data may be included. This chapter will explore the various types of filter conditions that can be employed in the where clauses of select, update, and delete statements.

Condition Evaluation

A where clause may contain one or more conditions, separated by the operators and and or. If there are multiple conditions separated only by the and operator, then all of the conditions must evaluate to true for the row to be included in the result set. Consider the following where clause:

```
WHERE title = 'Teller' AND start_date < '2003-01-01'
```

Given these two conditions, any employee who is either not a teller or began working for the bank in 2003 or later will be removed from consideration. While this example uses only two conditions, no matter how many conditions are in your where clause, if they are separated by the and operator then they must all evaluate to true for the row to be included in the result set.

If all conditions in the where clause are separated by the or operator, however, then only *one* of the conditions must evaluate to true for the row to be included in the result set. Consider the following two conditions:

```
WHERE title = 'Teller' OR start_date < '2003-01-01'
```

There are now various ways for an employee row to be included in the result set:

- The employee is a teller and was employed prior to 2003.
- The employee is a teller and was employed after January 1, 2003.
- The employee is something other than a teller but was employed prior to 2003.

Table 4-1 shows the possible outcomes for a where clause containing two conditions separated by the or operator.

Table 4-1. Two-condition evaluation using or

Intermediate result	Final result
WHERE true OR true	True
WHERE true OR false	True
WHERE false OR true	True
WHERE false OR false	False

In the case of the preceding example, the only way for a row to be excluded from the result set is if the employee is not a teller and was employed on or after January 1, 2003.

Using Parentheses

If your where clause includes three or more conditions using both the and and or operators, you should use parentheses to make clear your intent, both to the database server and to anyone else reading your code. Here's a where clause that extends the previous example by checking to make sure that the employee is still employed by the bank:

```
WHERE end_date IS NULL
  AND (title = 'Teller' OR start_date < '2003-01-01')
```

There are now three conditions; for a row to make it to the final result set, the first condition must evaluate to true, and either the second *or* third conditions (or both) must evaluate to true. Table 4-2 shows the possible outcomes for this where clause.

Table 4-2. Three-condition evaluation using and, or

Intermediate result	Final result
WHERE true AND (true OR true)	True
WHERE true AND (true OR false)	True

Table 4-2. Three-condition evaluation using and, or (continued)

Intermediate result	Final result
WHERE true AND (false OR true)	True
WHERE true AND (false OR false)	False
WHERE false AND (true OR true)	False
WHERE false AND (true OR false)	False
WHERE false AND (false OR true)	False
WHERE false AND (false OR false)	False

As you can see, the more conditions that you have in your where clause, the more combinations there are to be evaluated by the server. In this case, only three of the eight combinations yield a final result of true.

Using the not Operator

Hopefully, the previous three-condition example is fairly easy to understand. Consider the following modification, however:

```
WHERE end_date IS NULL
   AND NOT (title = 'Teller' OR start_date < '2003-01-01')
```

Did you spot the change from the previous example? I added the not operator after the and operator on the second line. Now, instead of looking for nonterminated employees who are either tellers or who began working for the bank prior to 2003, I am looking for nonterminated employees who are either nontellers or who began working for the bank in 2003 or later. Table 4-3 shows the possible outcomes for this example.

Table 4-3. Three-condition evaluation using and, or, and not

Intermediate result	Final result
WHERE true AND NOT (true OR true)	False
WHERE true AND NOT (true OR false)	False
WHERE true AND NOT (false OR true)	False
WHERE true AND NOT (false OR false)	True
WHERE false AND NOT (true OR true)	False
WHERE false AND NOT (true OR false)	False
WHERE false AND NOT (false OR true)	False
WHERE false AND NOT (false OR false)	False

While it is easy for the database server to handle, it is typically difficult for a person to evaluate a where clause that includes the not operator, which is why you won't encounter it very often. In this case, the where clause can be rewritten to avoid using the not operator:

```
WHERE end_date IS NULL
    AND (title != 'Teller' OR start_date >= '2003-01-01')
```

While I'm sure that the server doesn't have a preference, you probably have an easier time understanding this version of the where clause.

Building a Condition

Now that you have seen how multiple conditions are evaluated by the server, let's take a step back and look at what comprises a single condition. A condition is made up of one or more *expressions* coupled with one or more *operators*. An expression can be any of the following:

- A number
- A column in a table or view
- A string literal, such as 'Teller'
- A built-in function, such as CONCAT('Learning', ' ', 'SQL')
- A subquery
- A list of expressions, such as ('Teller', 'Head Teller', 'Operations Manager')

The operators used within conditions include:

- Comparison operators, such as =, !=, <, >, <>, LIKE, IN, and BETWEEN
- Arithmetic operators, such as +, -, *, and /

The following section demonstrates how these expressions and operators can be combined to manufacture the various types of conditions.

Condition Types

There are many different ways to filter out unwanted data. You can look for specific values, sets of values, or ranges of values to include or exclude, or you can use various pattern-searching techniques to look for partial matches when dealing with string data. The next four subsections will explore each of these condition types in detail.

Equality Conditions

A large percentage of the filter conditions that you write or come across will be of the form 'column = expression' as in:

```
title = 'Teller'
fed_id = '111-11-1111'
```

```
amount = 375.25
dept_id = (SELECT dept_id FROM department WHERE name = 'Loans')
```

Conditions such as these are called *equality conditions* because they equate one
expression to another. The first three examples equate a column to a literal (two
strings and a number), and the fourth example equates a column to the value
returned from a subquery. The following query uses two equality conditions; one in
the on clause (a join condition), and the other in the where clause (a filter condition):

```
mysql> SELECT pt.name product_type, p.name product
    -> FROM product p INNER JOIN product_type pt
    ->   ON p.product_type_cd = pt.product_type_cd
    -> WHERE pt.name = 'Customer Accounts';
+-------------------+------------------------+
| product_type      | product                |
+-------------------+------------------------+
| Customer Accounts | certificate of deposit |
| Customer Accounts | checking account       |
| Customer Accounts | money market account   |
| Customer Accounts | savings account        |
+-------------------+------------------------+
4 rows in set (0.08 sec)
```

This query shows all products that are customer account types.

Inequality conditions

Another fairly common type of condition is the *inequality condition*, which asserts
that two expressions are *not* equal. Here's the previous query with the filter condition in the where clause changed to an inequality condition:

```
mysql> SELECT pt.name product_type, p.name product
    -> FROM product p INNER JOIN product_type pt
    ->   ON p.product_type_cd = pt.product_type_cd
    -> WHERE pt.name != 'Customer Accounts';
+-----------------------------+------------------------+
| product_type                | product                |
+-----------------------------+------------------------+
| Individual and Business Loans | auto loan            |
| Individual and Business Loans | business line of credit |
| Individual and Business Loans | home mortgage        |
| Individual and Business Loans | small business loan  |
+-----------------------------+------------------------+
4 rows in set (0.00 sec)
```

This query shows all products that are *not* customer account types. When building
inequality conditions, you may choose to use either the != or <> operators.

Data modification using equality conditions

Equality/inequality conditions are commonly used when modifying data. For example, let's say that the bank has a policy of removing old account rows once per year.

Your task is to remove rows from the account table that were closed in 1999. Here's one way to tackle it:

```
DELETE FROM account
WHERE status = 'CLOSED' AND YEAR(close_date) = 1999;
```

This statement includes two equality conditions: one to find only closed accounts, and another to check for those accounts closed in 1999.

 When crafting examples of delete and update statements, I will try to write each statement such that no rows are modified. That way, when you execute the statements, your data will remain unchanged, and your output from select statements will always match that shown in this book.

Since MySQL sessions are in auto-commit mode by default (see Chapter 12), you would not have the ability to rollback (undo) any changes made to the example data if one of my statements modified the data. You, of course, can do whatever you want with the example data, including wiping it clean and rerunning the scripts I have provided, but I will try to leave it intact.

Range Conditions

Along with checking that an expression is equal to (or not equal to) another expression, you can build conditions that check if an expression falls within a certain range. This type of condition is common when working with numeric or temporal data. Consider the following query:

```
mysql> SELECT emp_id, fname, lname, start_date
    -> FROM employee
    -> WHERE start_date < '2003-01-01';
```

emp_id	fname	lname	start_date
1	Michael	Smith	2001-06-22
2	Susan	Barker	2002-09-12
3	Robert	Tyler	2000-02-09
4	Susan	Hawthorne	2002-04-24
8	Sarah	Parker	2002-12-02
9	Jane	Grossman	2002-05-03
10	Paula	Roberts	2002-07-27
11	Thomas	Ziegler	2000-10-23
13	John	Blake	2000-05-11
14	Cindy	Mason	2002-08-09
16	Theresa	Markham	2001-03-15
17	Beth	Fowler	2002-06-29
18	Rick	Tulman	2002-12-12

```
13 rows in set (0.01 sec)
```

This query finds all employees hired prior to 2003. Along with specifying an upper limit for the start date, you may also want to specify a lower range for the start date:

```
mysql> SELECT emp_id, fname, lname, start_date
    -> FROM employee
    -> WHERE start_date < '2003-01-01'
    ->   AND start_date >= '2001-01-01';
+--------+---------+-----------+------------+
| emp_id | fname   | lname     | start_date |
+--------+---------+-----------+------------+
|      1 | Michael | Smith     | 2001-06-22 |
|      2 | Susan   | Barker    | 2002-09-12 |
|      4 | Susan   | Hawthorne | 2002-04-24 |
|      8 | Sarah   | Parker    | 2002-12-02 |
|      9 | Jane    | Grossman  | 2002-05-03 |
|     10 | Paula   | Roberts   | 2002-07-27 |
|     14 | Cindy   | Mason     | 2002-08-09 |
|     16 | Theresa | Markham   | 2001-03-15 |
|     17 | Beth    | Fowler    | 2002-06-29 |
|     18 | Rick    | Tulman    | 2002-12-12 |
+--------+---------+-----------+------------+
10 rows in set (0.01 sec)
```

This version of the query retrieves all employees hired in 2001 or 2002.

The between operator

When you have *both* an upper and lower limit for your range, you may choose to use a single condition that utilizes the between operator rather than using two separate conditions, as in:

```
mysql> SELECT emp_id, fname, lname, start_date
    -> FROM employee
    -> WHERE start_date BETWEEN '2001-01-01' AND '2003-01-01';
+--------+---------+-----------+------------+
| emp_id | fname   | lname     | start_date |
+--------+---------+-----------+------------+
|      1 | Michael | Smith     | 2001-06-22 |
|      2 | Susan   | Barker    | 2002-09-12 |
|      4 | Susan   | Hawthorne | 2002-04-24 |
|      8 | Sarah   | Parker    | 2002-12-02 |
|      9 | Jane    | Grossman  | 2002-05-03 |
|     10 | Paula   | Roberts   | 2002-07-27 |
|     14 | Cindy   | Mason     | 2002-08-09 |
|     16 | Theresa | Markham   | 2001-03-15 |
|     17 | Beth    | Fowler    | 2002-06-29 |
|     18 | Rick    | Tulman    | 2002-12-12 |
+--------+---------+-----------+------------+
10 rows in set (0.05 sec)
```

When using the between operator, there are a couple of things to keep in mind. You should always specify the lower limit of the range first (after between) and the upper limit of the range second (after and). Here's what happens if you specify the upper limit first:

```
mysql> SELECT emp_id, fname, lname, start_date
    -> FROM employee
    -> WHERE start_date BETWEEN '2003-01-01' AND '2001-01-01';
Empty set (0.00 sec)
```

As you can see, no data is returned. This is because the server is, in effect, generating two conditions from your single condition using the <= and >= operators, as in:

```
mysql> SELECT emp_id, fname, lname, start_date
    -> FROM employee
    -> WHERE start_date >= '2003-01-01'
    ->   AND start_date <= '2001-01-01';
Empty set (0.00 sec)
```

Since it is impossible to have a date that is *both* greater than January 1, 2003, and less than January 1, 2001, the query returns an empty set. This brings me to the second pitfall when using between, which is to remember that your upper and lower limits are *inclusive*, meaning that the values you provide are included in the range limits. In this case, I want to specify 2001-01-01 as the lower end of the range and 2002-12-31 as the upper end, rather than 2003-01-01. Even though there probably weren't any employees who started working for the bank on New Year's Day, 2003, it is best to specify exactly what you want.

Along with dates, you can also build conditions to specify ranges of numbers. Numeric ranges are fairly easy to grasp, as demonstrated by the following:

```
mysql> SELECT account_id, product_cd, cust_id, avail_balance
    -> FROM account
    -> WHERE avail_balance BETWEEN 3000 AND 5000;
+------------+------------+---------+---------------+
| account_id | product_cd | cust_id | avail_balance |
+------------+------------+---------+---------------+
|          3 | CD         |       1 |       3000.00 |
|         14 | CD         |       7 |       5000.00 |
|         15 | CHK        |       8 |       3487.19 |
+------------+------------+---------+---------------+
3 rows in set (0.03 sec)
```

All accounts with between $3,000 and $5,000 of an available balance are returned. Again, make sure that you specify the lower amount first.

String ranges

While ranges of dates and numbers are easy to understand, you can also build conditions that search for ranges of strings, which are a bit harder to visualize. Say, for example, you are searching for customers having a Social Security number that falls within a certain range. The format for a Social Security number is "XXX-XX-XXXX," where X is a number from 0 to 9, and you want to find every customer whose Social

Security number lies between "500-00-0000" and "999-99-9999." Here's what the statement would look like:

```
mysql> SELECT cust_id, fed_id
    -> FROM customer
    -> WHERE cust_type_cd = 'I'
    ->   AND fed_id BETWEEN '500-00-0000' AND '999-99-9999';
+---------+-------------+
| cust_id | fed_id      |
+---------+-------------+
|       5 | 555-55-5555 |
|       6 | 666-66-6666 |
|       7 | 777-77-7777 |
|       8 | 888-88-8888 |
|       9 | 999-99-9999 |
+---------+-------------+
5 rows in set (0.01 sec)
```

To work with string ranges, you need to know the order of the characters within your character set (the order in which the characters within a character set are sorted is called a *collation*).

Membership Conditions

In some cases, you will not be restricting an expression to a single value or range of values, but rather to a finite set of values. For example, you might want to locate all accounts whose product code is either 'CHK', 'SAV', 'CD', or 'MM':

```
mysql> SELECT account_id, product_cd, cust_id, avail_balance
    -> FROM account
    -> WHERE product_cd = 'CHK' OR product_cd = 'SAV'
    ->   OR product_cd = 'CD' OR product_cd = 'MM';
+------------+------------+---------+---------------+
| account_id | product_cd | cust_id | avail_balance |
+------------+------------+---------+---------------+
|          1 | CHK        |       1 |       1057.75 |
|          2 | SAV        |       1 |        500.00 |
|          3 | CD         |       1 |       3000.00 |
|          4 | CHK        |       2 |       2258.02 |
|          5 | SAV        |       2 |        200.00 |
|          6 | CHK        |       3 |       1057.75 |
|          7 | MM         |       3 |       2212.50 |
|          8 | CHK        |       4 |        534.12 |
|          9 | SAV        |       4 |        767.77 |
|         10 | MM         |       4 |       5487.09 |
|         11 | CHK        |       5 |       2237.97 |
|         12 | CHK        |       6 |        122.37 |
|         13 | CD         |       6 |      10000.00 |
|         14 | CD         |       7 |       5000.00 |
|         15 | CHK        |       8 |       3487.19 |
|         16 | SAV        |       8 |        387.99 |
|         17 | CHK        |       9 |        125.67 |
|         18 | MM         |       9 |       9345.55 |
|         19 | CD         |       9 |       1500.00 |
```

```
|          20 | CHK        |      10 |       23575.12 |
|          23 | CHK        |      12 |       38552.05 |
+-------------+------------+---------+----------------+
21 rows in set (0.02 sec)
```

While this where clause (four conditions or'd together) wasn't too tedious to gencr-
ate, imagine if the set of expressions contained 10 or 20 members? For these situations, you can use the in operator instead:

```
SELECT account_id, product_cd, cust_id, avail_balance
FROM account
WHERE product_cd IN ('CHK','SAV','CD','MM');
```

With the in operator, you can write a single condition no matter how many expressions are in the set.

Using subqueries

Along with writing your own set of expressions, such as ('CHK','SAV','CD','MM'),
you can use a subquery to generate a set for you. For example, all four of the product types used in the previous query have a product_type_cd of 'ACCOUNT'. The next
version of the query uses a subquery against the product table to retrieve the four
product codes instead of explicitly naming them:

```
mysql> SELECT account_id, product_cd, cust_id, avail_balance
    ->   FROM account
    ->   WHERE product_cd IN (SELECT product_cd FROM product
    ->     WHERE product_type_cd = 'ACCOUNT');
+-------------+------------+---------+----------------+
| account_id  | product_cd | cust_id | avail_balance  |
+-------------+------------+---------+----------------+
|           1 | CHK        |       1 |        1057.75 |
|           2 | SAV        |       1 |         500.00 |
|           3 | CD         |       1 |        3000.00 |
|           4 | CHK        |       2 |        2258.02 |
|           5 | SAV        |       2 |         200.00 |
|           6 | CHK        |       3 |        1057.75 |
|           7 | MM         |       3 |        2212.50 |
|           8 | CHK        |       4 |         534.12 |
|           9 | SAV        |       4 |         767.77 |
|          10 | MM         |       4 |        5487.09 |
|          11 | CHK        |       5 |        2237.97 |
|          12 | CHK        |       6 |         122.37 |
|          13 | CD         |       6 |       10000.00 |
|          14 | CD         |       7 |        5000.00 |
|          15 | CHK        |       8 |        3487.19 |
|          16 | SAV        |       8 |         387.99 |
|          17 | CHK        |       9 |         125.67 |
|          18 | MM         |       9 |        9345.55 |
|          19 | CD         |       9 |        1500.00 |
|          20 | CHK        |      10 |       23575.12 |
|          23 | CHK        |      12 |       38552.05 |
+-------------+------------+---------+----------------+
21 rows in set (0.03 sec)
```

The subquery returns a set of four values, and the main query checks to see if the value of the product_cd column can be found in the set returned by the subquery.

Using not in

Sometimes you want to see if a particular expression exists within a set of expressions, and sometimes you want to see if the expression does *not* exist. For these situations, you can use the not in operator:

```
mysql> SELECT account_id, product_cd, cust_id, avail_balance
    ->    FROM account
    ->    WHERE product_cd NOT IN ('CHK','SAV','CD','MM');
+------------+------------+---------+---------------+
| account_id | product_cd | cust_id | avail_balance |
+------------+------------+---------+---------------+
|         21 | BUS        |      10 |          0.00 |
|         22 | BUS        |      11 |       9345.55 |
|         24 | SBL        |      13 |      50000.00 |
+------------+------------+---------+---------------+
3 rows in set (0.02 sec)
```

This query finds all accounts that are *not* checking, savings, certificate of deposit, or money market accounts.

Matching Conditions

So far, you have been introduced to conditions that identify an exact string, a range of strings, or a set of strings; the final condition type deals with partial string matches. You may, for example, want to find all employees whose last name begins with "T." You could use a built-in function to strip off the first letter of the lname column, as in:

```
mysql> SELECT emp_id, fname, lname
    -> FROM employee
    -> WHERE LEFT(lname, 1) = 'T';
+--------+--------+--------+
| emp_id | fname  | lname  |
+--------+--------+--------+
|      3 | Robert | Tyler  |
|      7 | Chris  | Tucker |
|     18 | Rick   | Tulman |
+--------+--------+--------+
3 rows in set (0.01 sec)
```

While the built-in function left( ) does the job, it doesn't give you much flexibility. Instead, you can use wildcard characters to build search expressions, as demonstrated in the next section.

Using wildcards

When searching for partial string matches, you might be interested in:

- Strings beginning/ending with a certain character
- Strings beginning/ending with a substring
- Strings containing a certain character anywhere within the string
- Strings containing a substring anywhere within the string
- Strings with a specific format, regardless of individual characters

You can build search expressions to identify these and many other partial string matches by using the wildcard characters shown in Table 4-4.

Table 4-4. Wildcard characters

Wildcard character	Matches
_	Exactly one character
%	Any number of characters (including zero)

The underscore character takes the place of a single character, while the percent sign can take the place of a variable number of characters. When building conditions that utilize search expressions, you use the like operator, as in:

```
mysql> SELECT lname
    -> FROM employee
    -> WHERE lname LIKE '_a%e%';
+-----------+
| lname     |
+-----------+
| Barker    |
| Hawthorne |
| Parker    |
| Jameson   |
+-----------+
4 rows in set (0.00 sec)
```

The search expression in the previous example specifies strings containing an "a" in the second position and followed by an "e" at any other position in the string (including the last position). Table 4-5 shows some more search expressions and their interpretations.

Table 4-5. Sample search expressions

Search expression	Interpretation
F%	Strings beginning with "F"
%t	Strings ending with "t"
%bas%	Strings containing the substring "bas"

Table 4-5. Sample search expressions (continued)

Search expression	Interpretation
__t_	Four-character strings with a "t" in the third position
___-__-____	11-character strings with dashes in the fourth and seventh positions

The last example in Table 4-5 could be used to find customers whose Federal ID matches the format used for Social Security numbers, as in:

```
mysql> SELECT cust_id, fed_id
    -> FROM customer
    -> WHERE fed_id LIKE '___-__-____';
+---------+-------------+
| cust_id | fed_id      |
+---------+-------------+
|       1 | 111-11-1111 |
|       2 | 222-22-2222 |
|       3 | 333-33-3333 |
|       4 | 444-44-4444 |
|       5 | 555-55-5555 |
|       6 | 666-66-6666 |
|       7 | 777-77-7777 |
|       8 | 888-88-8888 |
|       9 | 999-99-9999 |
+---------+-------------+
9 rows in set (0.02 sec)
```

The wildcard characters work fine for building simple search expressions; if your needs are a bit more sophisticated, you can use multiple search expressions, as demonstrated by the following:

```
mysql> SELECT emp_id, fname, lname
    -> FROM employee
    -> WHERE lname LIKE 'F%' OR lname LIKE 'G%';
+--------+-------+----------+
| emp_id | fname | lname    |
+--------+-------+----------+
|      5 | John  | Gooding  |
|      6 | Helen | Fleming  |
|      9 | Jane  | Grossman |
|     17 | Beth  | Fowler   |
+--------+-------+----------+
4 rows in set (0.00 sec)
```

This query finds all employees whose last name begins with "F" or "G."

Using regular expressions

If you find that the wildcard characters don't provide enough flexibility, you can use regular expressions to build search expressions. A regular expression is, in essence, a search expression on steroids. If you are new to SQL but have coded using programming languages such as Perl, then you might already be intimately familiar with regular expressions. If you have never used regular expressions, then you may want to consult *Mastering Regular Expressions* by Jeffrey Friedl (O'Reilly), since it is far too large a topic to try to cover in this book.

Here's what the previous query (find all employees whose last name starts with "F" or "G") would look like using the MySQL implementation of regular expressions:

```
mysql> SELECT emp_id, fname, lname
    -> FROM employee
    -> WHERE lname REGEXP '^[FG]';
+--------+-------+----------+
| emp_id | fname | lname    |
+--------+-------+----------+
|      5 | John  | Gooding  |
|      6 | Helen | Fleming  |
|      9 | Jane  | Grossman |
|     17 | Beth  | Fowler   |
+--------+-------+----------+
4 rows in set (0.00 sec)
```

The regexp operator takes a regular expression ('^[FG]' in this example) and applies it to the expression on the lefthand side of the condition (the column lname). The query now contains a single condition using a regular expression rather than two conditions using wildcard characters.

Oracle Database 10g and SQL Server 2000 also support regular expressions. With Oracle Database 10g, you would use the regexp_like function instead of the regexp operator shown in the previous example, whereas SQL Server allows regular expressions to be used with the like operator.

NULL: That Four-Letter Word

I've put it off as long as I could, but it's time to broach a topic that tends to be met with fear, uncertainty, and dread: the null value. null is the absence of a value; before an employee is terminated, for example, her end_date column in the employee table should be null. There is no value that can be assigned to the end_date column that would make sense in this situation. Null is a bit slippery, however, as there are various flavors of null:

Not applicable
Such as the employee ID column for a transaction that took place at an ATM machine

Value not yet known
Such as when the Federal ID is not known at the time a customer row is created

Value undefined
Such as when an account is created for a product that has not yet been added to the database

Some theorists argue that there should be a different expression to cover each of these (and more) situations, but most practitioners would agree that having multiple null values would be far too confusing.

When working with null, you should remember:

- An expression can *be* null, but it can never *equal* null.
- Two nulls are never equal to each other.

To test whether an expression is null, you need to use the is null operator, as demonstrated by the following:

```
mysql> SELECT emp_id, fname, lname, superior_emp_id
    -> FROM employee
    -> WHERE superior_emp_id IS NULL;
+--------+---------+-------+-----------------+
| emp_id | fname   | lname | superior_emp_id |
+--------+---------+-------+-----------------+
|      1 | Michael | Smith |            NULL |
+--------+---------+-------+-----------------+
1 row in set (0.00 sec)
```

This query returns all employees who do not have a boss. Here's the same query using = null instead of is null:

```
mysql> SELECT emp_id, fname, lname, superior_emp_id
    -> FROM employee
    -> WHERE superior_emp_id = NULL;
Empty set (0.01 sec)
```

As you can see, the query parses and executes but does not return any rows. This is a common mistake made by inexperienced SQL programmers, and the database server will not alert you to your error, so be careful when constructing conditions that test for null.

If you want to see if a value has been assigned to a column, you can use the is not null operator, as in:

```
mysql> SELECT emp_id, fname, lname, superior_emp_id
    -> FROM employee
    -> WHERE superior_emp_id IS NOT NULL;
+--------+---------+-----------+-----------------+
| emp_id | fname   | lname     | superior_emp_id |
+--------+---------+-----------+-----------------+
|      2 | Susan   | Barker    |               1 |
|      3 | Robert  | Tyler     |               1 |
|      4 | Susan   | Hawthorne |               3 |
|      5 | John    | Gooding   |               4 |
|      6 | Helen   | Fleming   |               4 |
|      7 | Chris   | Tucker    |               6 |
|      8 | Sarah   | Parker    |               6 |
|      9 | Jane    | Grossman  |               6 |
```

```
|     10 | Paula    | Roberts   |                4 |
|     11 | Thomas   | Ziegler   |               10 |
|     12 | Samantha | Jameson   |               10 |
|     13 | John     | Blake     |                4 |
|     14 | Cindy    | Mason     |               13 |
|     15 | Frank    | Portman   |               13 |
|     16 | Theresa  | Markham   |                4 |
|     17 | Beth     | Fowler    |               16 |
|     18 | Rick     | Tulman    |               16 |
+--------+----------+-----------+------------------+
17 rows in set (0.01 sec)
```

This version of the query returns the other 17 employees who, unlike Michael Smith, have a boss.

Before putting null aside for awhile, it would be helpful to investigate one more potential pitfall. Suppose that you have been asked to identify all employees who are not managed by Helen Fleming, whose employee ID is 6. Your first instinct might be to do the following:

```
mysql> SELECT emp_id, fname, lname, superior_emp_id
    -> FROM employee
    -> WHERE superior_emp_id != 6;
+--------+----------+-----------+------------------+
| emp_id | fname    | lname     | superior_emp_id  |
+--------+----------+-----------+------------------+
|      2 | Susan    | Barker    |                1 |
|      3 | Robert   | Tyler     |                1 |
|      4 | Susan    | Hawthorne |                3 |
|      5 | John     | Gooding   |                4 |
|      6 | Helen    | Fleming   |                4 |
|     10 | Paula    | Roberts   |                4 |
|     11 | Thomas   | Ziegler   |               10 |
|     12 | Samantha | Jameson   |               10 |
|     13 | John     | Blake     |                4 |
|     14 | Cindy    | Mason     |               13 |
|     15 | Frank    | Portman   |               13 |
|     16 | Theresa  | Markham   |                4 |
|     17 | Beth     | Fowler    |               16 |
|     18 | Rick     | Tulman    |               16 |
+--------+----------+-----------+------------------+
14 rows in set (0.01 sec)
```

While it is true that these 14 employees do not work for Helen Fleming, if you look carefully at the data you will see that there is one more employee who doesn't work for Helen who is not listed here. That employee is Michael Smith, and his superior_emp_id column is null (because he's the big cheese). To answer the question correctly, therefore, you need to account for the possibility that some rows might contain a null in the superior_emp_id column:

```
mysql> SELECT emp_id, fname, lname, superior_emp_id
    -> FROM employee
    -> WHERE superior_emp_id != 6 OR superior_emp_id IS NULL;
```

```
+---------+-----------+-----------+------------------+
| emp_id  | fname     | lname     | superior_emp_id  |
+---------+-----------+-----------+------------------+
|       1 | Michael   | Smith     |             NULL |
|       2 | Susan     | Barker    |                1 |
|       3 | Robert    | Tyler     |                1 |
|       4 | Susan     | Hawthorne |                3 |
|       5 | John      | Gooding   |                4 |
|       6 | Helen     | Fleming   |                4 |
|      10 | Paula     | Roberts   |                4 |
|      11 | Thomas    | Ziegler   |               10 |
|      12 | Samantha  | Jameson   |               10 |
|      13 | John      | Blake     |                4 |
|      14 | Cindy     | Mason     |               13 |
|      15 | Frank     | Portman   |               13 |
|      16 | Theresa   | Markham   |                4 |
|      17 | Beth      | Fowler    |               16 |
|      18 | Rick      | Tulman    |               16 |
+---------+-----------+-----------+------------------+
15 rows in set (0.01 sec)
```

The result set now includes all 15 employees who don't work for Helen. When working with a database that you are not familiar with, it is a good idea to find out which columns in a table allow nulls so that you can take appropriate measures with your filter conditions to keep data from slipping through the cracks.

Exercises

The following exercises will test your understanding of filter conditions. Please see Appendix C for solutions.

The following transaction data is used for the first two exercises:

Txn_id	Txn_date	Account_id	Txn_type_cd	Amount
1	2005-02-22	101	CDT	1000.00
2	2005-02-23	102	CDT	525.75
3	2005-02-24	101	DBT	100.00
4	2005-02-24	103	CDT	55
5	2005-02-25	101	DBT	50
6	2005-02-25	103	DBT	25
7	2005-02-25	102	CDT	125.37
8	2005-02-26	103	DBT	10
9	2005-02-27	101	CDT	75

4-1

Which of the transaction IDs would be returned by the following filter conditions?

```
txn_date < '2005-02-26' AND (txn_type_cd = 'DBT' OR amount > 100)
```

4-2

Which of the transaction IDs would be returned by the following filter conditions?

```
account_id IN (101,103) AND NOT (txn_type_cd = 'DBT' OR amount > 100)
```

4-3

Construct a query that retrieves all accounts opened in 2002.

4-4

Construct a query that finds all nonbusiness customers whose last name contains an 'a' in the second position and an 'e' anywhere after the 'a'.

CHAPTER 5
Querying Multiple Tables

Because relational database design mandates that independent entities be placed in separate tables, you will need a mechanism for bringing multiple tables together in the same query. This mechanism is known as a *join*, and this chapter will concentrate on the simplest and most common join, the *inner join*; Chapter 10 will demonstrate all of the different join types.

What Is a Join?

Queries against a single table are certainly not rare, but you will find that most of your queries will require two, three, or even more tables. To illustrate, let's look at the definitions for the employee and department tables and then define a query that retrieves data from both tables:

```
mysql> DESC employee;
+-------------------+----------------------+------+-----+------------+
| Field             | Type                 | Null | Key | Default    |
+-------------------+----------------------+------+-----+------------+
| emp_id            | smallint(5) unsigned |      | PRI | NULL       |
| fname             | varchar(20)          |      |     |            |
| lname             | varchar(20)          |      |     |            |
| start_date        | date                 |      |     | 0000-00-00 |
| end_date          | date                 | YES  |     | NULL       |
| superior_emp_id   | smallint(5) unsigned | YES  | MUL | NULL       |
| dept_id           | smallint(5) unsigned | YES  | MUL | NULL       |
| title             | varchar(20)          | YES  |     | NULL       |
| assigned_branch_id| smallint(5) unsigned | YES  | MUL | NULL       |
+-------------------+----------------------+------+-----+------------+
9 rows in set (0.11 sec)

mysql> DESC department;
+---------+----------------------+------+-----+---------+
| Field   | Type                 | Null | Key | Default |
+---------+----------------------+------+-----+---------+
| dept_id | smallint(5) unsigned |      | PRI | NULL    |
```

```
| name     | varchar(20)        |       |     |          |
+---------+-------------------+------+-----+---------+
2 rows in set (0.03 sec)
```

Let's say you want to retrieve the first and last names of each employee along with the name of the department to which each employee is assigned. Your query will therefore need to retrieve the `employee.fname`, `employee.lname`, and `department.name` columns. But how can you retrieve data from both tables in the same query? The answer lies in the `employee.dept_id` column, which holds the ID of the department to which each employee is assigned (in more formal terms, the `employee.dept_id` column is the *foreign key* to the department table). The query, which you will see shortly, instructs the server to use the `employee.dept_id` column as the *bridge* between the employee and department tables, thereby allowing columns from both tables to be included in the query's result set. This type of operation is known as a join.

Cartesian Product

The easiest way to start is to put the `employee` and `department` tables into the `from` clause of a query and see what happens. Here's a query that retrieves the employee's first and last names along with the department name, with a `from` clause naming both tables separated by the join keyword:

```
mysql> SELECT e.fname, e.lname, d.name
    -> FROM employee e JOIN department d;
+----------+----------+-----------------+
| fname    | lname    | name            |
+----------+----------+-----------------+
| Michael  | Smith    | Operations      |
| Susan    | Barker   | Operations      |
| Robert   | Tyler    | Operations      |
| Susan    | Hawthorne| Operations      |
| John     | Gooding  | Operations      |
| Helen    | Fleming  | Operations      |
| Chris    | Tucker   | Operations      |
| Sarah    | Parker   | Operations      |
| Jane     | Grossman | Operations      |
| Paula    | Roberts  | Operations      |
| Thomas   | Ziegler  | Operations      |
| Samantha | Jameson  | Operations      |
| John     | Blake    | Operations      |
| Cindy    | Mason    | Operations      |
| Frank    | Portman  | Operations      |
| Theresa  | Markham  | Operations      |
| Beth     | Fowler   | Operations      |
| Rick     | Tulman   | Operations      |
| Michael  | Smith    | Loans           |
| Susan    | Barker   | Loans           |
| Robert   | Tyler    | Loans           |
| Susan    | Hawthorne| Loans           |
```

```
| John     | Gooding   | Loans          |
| Helen    | Fleming   | Loans          |
| Chris    | Tucker    | Loans          |
| Sarah    | Parker    | Loans          |
| Jane     | Grossman  | Loans          |
| Paula    | Roberts   | Loans          |
| Thomas   | Ziegler   | Loans          |
| Samantha | Jameson   | Loans          |
| John     | Blake     | Loans          |
| Cindy    | Mason     | Loans          |
| Frank    | Portman   | Loans          |
| Theresa  | Markham   | Loans          |
| Beth     | Fowler    | Loans          |
| Rick     | Tulman    | Loans          |
| Michael  | Smith     | Administration |
| Susan    | Barker    | Administration |
| Robert   | Tyler     | Administration |
| Susan    | Hawthorne | Administration |
| John     | Gooding   | Administration |
| Helen    | Fleming   | Administration |
| Chris    | Tucker    | Administration |
| Sarah    | Parker    | Administration |
| Jane     | Grossman  | Administration |
| Paula    | Roberts   | Administration |
| Thomas   | Ziegler   | Administration |
| Samantha | Jameson   | Administration |
| John     | Blake     | Administration |
| Cindy    | Mason     | Administration |
| Frank    | Portman   | Administration |
| Theresa  | Markham   | Administration |
| Beth     | Fowler    | Administration |
| Rick     | Tulman    | Administration |
+----------+-----------+----------------+
54 rows in set (0.00 sec)
```

Hmmm…there are only 18 employees and 3 different departments, so how did the result set end up with 54 rows? Looking more closely, you can see that the set of 18 employees is repeated three times, with all of the data identical except for the department name. Because the query didn't specify *how* the two tables should be joined, the database server generated the *Cartesian product*, which is *every* permutation of the two tables (18 employees times 3 departments equals 54 permutations). This type of join is known as a *cross join*, and it is rarely used (on purpose, at least). Cross joins are one of the join types that will be studied in Chapter 10.

Inner Joins

To modify the previous query so that only 18 rows are included in the result set (one for each employee), you will need to describe how the two tables are related. Earlier,

I showed that the employee.dept_id column serves as the link between the two tables, so this information needs to be added to the on subclause of the from clause:

```
mysql> SELECT e.fname, e.lname, d.name
    -> FROM employee e JOIN department d
    ->   ON e.dept_id = d.dept_id;
+----------+-----------+----------------+
| fname    | lname     | name           |
+----------+-----------+----------------+
| Susan    | Hawthorne | Operations     |
| Helen    | Fleming   | Operations     |
| Chris    | Tucker    | Operations     |
| Sarah    | Parker    | Operations     |
| Jane     | Grossman  | Operations     |
| Paula    | Roberts   | Operations     |
| Thomas   | Ziegler   | Operations     |
| Samantha | Jameson   | Operations     |
| John     | Blake     | Operations     |
| Cindy    | Mason     | Operations     |
| Frank    | Portman   | Operations     |
| Theresa  | Markham   | Operations     |
| Beth     | Fowler    | Operations     |
| Rick     | Tulman    | Operations     |
| John     | Gooding   | Loans          |
| Michael  | Smith     | Administration |
| Susan    | Barker    | Administration |
| Robert   | Tyler     | Administration |
+----------+-----------+----------------+
18 rows in set (0.00 sec)
```

Instead of 54 rows, you now have the expected 18 rows due to the addition of the on subclause, which instructs the server to join the employee and department tables by using the dept_id column to traverse from one table to the other. For example, Susan Hawthorne's row in the employee table contains a value of 1 in the dept_id column (not shown in the example). The server uses this value to look up the row in the department table having a value of 1 in it's dept_id column and then retrieves the value 'Operations' from the name column in that row.

If a value exists for the dept_id column in one table but *not* the other, then the join fails for the rows containing that value and those rows are excluded from the result set. This type of join is known as the *inner join*, and it is the most commonly used type of join. To clarify, if the department table contains a fourth row for the marketing department, but no employees have been assigned to that department, then the marketing department would not be included in the result set. Likewise, if some of the employees had been assigned to department ID 99, which doesn't exist in the department table, then these employees would be left out of the result set. If you want to include all rows from one table or the other regardless of whether a match exists, you can specify an *outer join*, but this will be covered in Chapter 10.

In the previous example, I did not specify in the from clause which type of join to use. However, when you wish to join two tables using an inner join, you should explicitly specify this in your from clause; here's the same example, with the addition of the join type (note the keyword INNER):

```
mysql> SELECT e.fname, e.lname, d.name
    -> FROM employee e INNER JOIN department d
    ->   ON e.dept_id = d.dept_id;
+----------+-----------+----------------+
| fname    | lname     | name           |
+----------+-----------+----------------+
| Susan    | Hawthorne | Operations     |
| Helen    | Fleming   | Operations     |
| Chris    | Tucker    | Operations     |
| Sarah    | Parker    | Operations     |
| Jane     | Grossman  | Operations     |
| Paula    | Roberts   | Operations     |
| Thomas   | Ziegler   | Operations     |
| Samantha | Jameson   | Operations     |
| John     | Blake     | Operations     |
| Cindy    | Mason     | Operations     |
| Frank    | Portman   | Operations     |
| Theresa  | Markham   | Operations     |
| Beth     | Fowler    | Operations     |
| Rick     | Tulman    | Operations     |
| John     | Gooding   | Loans          |
| Michael  | Smith     | Administration |
| Susan    | Barker    | Administration |
| Robert   | Tyler     | Administration |
+----------+-----------+----------------+
18 rows in set (0.00 sec)
```

If you do not specify the type of join, then the server will do an inner join by default. As you will see in Chapter 10, however, there are several types of joins, so you should get in the habit of specifying the exact type of join that you require.

If the names of the columns used to join the two tables are identical, which is true in the previous query, you can use the using subclause instead of the on subclause, as in:

```
mysql> SELECT e.fname, e.lname, d.name
    -> FROM employee e INNER JOIN department d
    ->   USING (dept_id);
+----------+-----------+----------------+
| fname    | lname     | name           |
+----------+-----------+----------------+
| Susan    | Hawthorne | Operations     |
| Helen    | Fleming   | Operations     |
| Chris    | Tucker    | Operations     |
| Sarah    | Parker    | Operations     |
| Jane     | Grossman  | Operations     |
| Paula    | Roberts   | Operations     |
| Thomas   | Ziegler   | Operations     |
| Samantha | Jameson   | Operations     |
```

```
| John     | Blake    | Operations     |
| Cindy    | Mason    | Operations     |
| Frank    | Portman  | Operations     |
| Theresa  | Markham  | Operations     |
| Beth     | Fowler   | Operations     |
| Rick     | Tulman   | Operations     |
| John     | Gooding  | Loans          |
| Michael  | Smith    | Administration |
| Susan    | Barker   | Administration |
| Robert   | Tyler    | Administration |
+----------+----------+----------------+
18 rows in set (0.01 sec)
```

Since using is a shorthand notation that can be used only in a specific situation, I prefer always to use the on subclause to avoid confusion.

The ANSI Join Syntax

The notation used throughout this book for joining tables was introduced in the SQL92 version of the ANSI SQL standard. All of the major databases (Oracle Database, Microsoft SQL Server, MySQL, IBM DB2 Universal Database, Sybase Adaptive Server) have adopted the SQL92 join syntax. Because most of these servers have been around since before the release of the SQL92 specification, they all include an older join syntax as well. For example, all of these servers would understand the following variation of the previous query:

```
mysql> SELECT e.fname, e.lname, d.name
    -> FROM employee e, department d
    -> WHERE e.dept_id = d.dept_id;
+----------+----------+----------------+
| fname    | lname    | name           |
+----------+----------+----------------+
| Susan    | Hawthorne| Operations     |
| Helen    | Fleming  | Operations     |
| Chris    | Tucker   | Operations     |
| Sarah    | Parker   | Operations     |
| Jane     | Grossman | Operations     |
| Paula    | Roberts  | Operations     |
| Thomas   | Ziegler  | Operations     |
| Samantha | Jameson  | Operations     |
| John     | Blake    | Operations     |
| Cindy    | Mason    | Operations     |
| Frank    | Portman  | Operations     |
| Theresa  | Markham  | Operations     |
| Beth     | Fowler   | Operations     |
| Rick     | Tulman   | Operations     |
| John     | Gooding  | Loans          |
| Michael  | Smith    | Administration |
| Susan    | Barker   | Administration |
| Robert   | Tyler    | Administration |
+----------+----------+----------------+
18 rows in set (0.01 sec)
```

This older method of specifying joins does not include the on subclause; instead, tables are named in the from clause separated by commas, and join conditions are included in the where clause. While you may decide to ignore the SQL92 syntax in favor of the older join syntax, the ANSI join syntax has the following advantages:

- Join conditions and filter conditions are separated into two different clauses (the on subclause and the where clause, respectively), making a query easier to understand.

- The join conditions for each pair of tables is contained in its own on clause, making it less likely that part of a join will be mistakenly excluded.

- Queries that use the SQL92 join syntax are portable across database servers, whereas the older syntax is slightly different across the different servers.

The benefits of the SQL92 join syntax are easier to identify for complex queries that include both join and filter conditions. Consider the following query, which returns all accounts opened by experienced tellers (hired prior to 2003) currently assigned to the Woburn branch:

```
mysql> SELECT a.account_id, a.cust_id, a.open_date, a.product_cd
    -> FROM account a, branch b, employee e
    -> WHERE a.open_emp_id = e.emp_id
    ->   AND e.start_date <= '2003-01-01'
    ->   AND e.assigned_branch_id = b.branch_id
    ->   AND (e.title = 'Teller' OR e.title = 'Head Teller')
    ->   AND b.name = 'Woburn Branch';
+------------+---------+------------+------------+
| account_id | cust_id | open_date  | product_cd |
+------------+---------+------------+------------+
|          1 |       1 | 2000-01-15 | CHK        |
|          2 |       1 | 2000-01-15 | SAV        |
|          3 |       1 | 2004-06-30 | CD         |
|          4 |       2 | 2001-03-12 | CHK        |
|          5 |       2 | 2001-03-12 | SAV        |
|         14 |       7 | 2004-01-12 | CD         |
|         22 |      11 | 2004-03-22 | BUS        |
+------------+---------+------------+------------+
7 rows in set (0.01 sec)
```

With this query, it is not so easy to determine which conditions in the where clause are join conditions and which are filter conditions. It is also not readily apparent which type of join is being employed (to identify the type of join, you would need to look closely at the join conditions in the where clause to see if any special characters are employed), nor is it easy to determine whether any join conditions have been mistakenly left out. Here's the same query using the SQL92 join syntax:

```
mysql> SELECT a.account_id, a.cust_id, a.open_date, a.product_cd
    -> FROM account a INNER JOIN employee e
    ->   ON a.open_emp_id = e.emp_id
    -> INNER JOIN branch b
    ->   ON e.assigned_branch_id = b.branch_id
```

```
    -> WHERE e.start_date <= '2003-01-01'
    ->    AND (e.title = 'Teller' OR e.title = 'Head Teller')
    ->    AND b.name = 'Woburn Branch';
+------------+---------+------------+------------+
| account_id | cust_id | open_date  | product_cd |
+------------+---------+------------+------------+
|          1 |       1 | 2000-01-15 | CHK        |
|          2 |       1 | 2000-01-15 | SAV        |
|          3 |       1 | 2004-06-30 | CD         |
|          4 |       2 | 2001-03-12 | CHK        |
|          5 |       2 | 2001-03-12 | SAV        |
|         14 |       7 | 2004-01-12 | CD         |
|         22 |      11 | 2004-03-22 | BUS        |
+------------+---------+------------+------------+
7 rows in set (0.36 sec)
```

Hopefully, you agree that the version using SQL92 join syntax is easier to understand.

Joining Three or More Tables

Joining three tables is similar to joining two tables, but with one slight wrinkle. With a two-table join, there are two tables and one join type in the from clause, and a single on subclause to define how the tables are joined. With a three-table join, there are three tables and two join types in the from clause and two on subclauses. Here's another example of a query with a two-table join:

```
mysql> SELECT a.account_id, c.fed_id
    -> FROM account a INNER JOIN customer c
    ->   ON a.cust_id = c.cust_id
    -> WHERE c.cust_type_cd = 'B';
+------------+------------+
| account_id | fed_id     |
+------------+------------+
|         20 | 04-1111111 |
|         21 | 04-1111111 |
|         22 | 04-2222222 |
|         23 | 04-3333333 |
|         24 | 04-4444444 |
+------------+------------+
5 rows in set (0.06 sec)
```

This query, which returns the account ID and federal tax number for all business accounts, should look fairly straightforward by now. If, however, you add the employee table to the query to retrieve the name of the teller who opened each account, it looks as follows:

```
mysql> SELECT a.account_id, c.fed_id, e.fname, e.lname
    -> FROM account a INNER JOIN customer c
    ->   ON a.cust_id = c.cust_id
    ->   INNER JOIN employee e
    ->   ON a.open_emp_id = e.emp_id
    -> WHERE c.cust_type_cd = 'B';
```

```
+------------+------------+---------+---------+
| account_id | fed_id     | fname   | lname   |
+------------+------------+---------+---------+
|         20 | 04-1111111 | Theresa | Markham |
|         21 | 04-1111111 | Theresa | Markham |
|         22 | 04-2222222 | Paula   | Roberts |
|         23 | 04-3333333 | Theresa | Markham |
|         24 | 04-4444444 | John    | Blake   |
+------------+------------+---------+---------+
5 rows in set (0.03 sec)
```

There are now three tables, two join types, and two on subclauses listed in the from clause, so things have gotten quite a bit busier. At first glance, the order in which the tables are named might cause you to think that the employee table is being joined to the customer table, since the account table is named first, followed by the customer table, and then the employee table. If you switch the order in which the first two tables appear, however, you will get the exact same results:

```
mysql> SELECT a.account_id, c.fed_id, e.fname, e.lname
    -> FROM customer c INNER JOIN account a
    ->   ON a.cust_id = c.cust_id
    ->   INNER JOIN employee e
    ->   ON a.open_emp_id = e.emp_id
    -> WHERE c.cust_type_cd = 'B';
+------------+------------+---------+---------+
| account_id | fed_id     | fname   | lname   |
+------------+------------+---------+---------+
|         20 | 04-1111111 | Theresa | Markham |
|         21 | 04-1111111 | Theresa | Markham |
|         22 | 04-2222222 | Paula   | Roberts |
|         23 | 04-3333333 | Theresa | Markham |
|         24 | 04-4444444 | John    | Blake   |
+------------+------------+---------+---------+
5 rows in set (0.00 sec)
```

The customer table is now listed first, followed by the account table and then the employee table. Since the on subclauses haven't changed, the results are the same.

One way to think of a query that uses three or more tables is as a snowball rolling down a hill. The first two tables get the ball rolling, and each subsequent table gets tacked on to the snowball as it heads downhill. You can think of the snowball as the *intermediate result set*, which is picking up more and more columns as subsequent tables are joined. Therefore, the employee table is not really being joined to the account table, but rather the intermediate result set created when the customer and account tables were joined. (In case you were wondering why I chose a snowball analogy, I wrote this chapter in the midst of a New England winter: 110 inches so far, and more coming tomorrow. Oh joy.)

Using Subqueries as Tables

You have already seen several examples of queries that use three tables, but there is one variation worth mentioning: what to do if some of the tables are generated by subqueries. Subqueries will be the focus of Chapter 9, but I have already introduced the concept of a subquery in the from clause in the previous chapter. Here's another version of an earlier query (find all accounts opened by experienced tellers currently assigned to the Woburn branch) that joins the account table to subqueries against the branch and employee tables:

```
1  SELECT a.account_id, a.cust_id, a.open_date, a.product_cd
2  FROM account a INNER JOIN
3    (SELECT emp_id, assigned_branch_id
4     FROM employee
5     WHERE start_date <= '2003-01-01'
6       AND (title = 'Teller' OR title = 'Head Teller')) e
7  ON a.open_emp_id = e.emp_id
8  INNER JOIN
9    (SELECT branch_id
10    FROM branch
11    WHERE name = 'Woburn Branch') b
12 ON e.assigned_branch_id = b.branch_id;
```

The first subquery, which starts on line 3 and is given the alias e, finds all inexperienced tellers. The second subquery, which starts on line 9 and is given the alias b, finds the ID of the Woburn branch. First, the account table is joined to the inexperienced teller subquery using the employee ID, and then the resulting table is joined to the Woburn branch subquery using the branch ID. The results are the same as those of the previous version of the query (try it and see for yourself), but the queries look very different from one another.

There isn't really anything shocking here, but it might take a minute to figure out what's going on. Notice, for example, the lack of a where clause in the main query; since all of the filter conditions are against the employee and branch tables, the filter conditions are all inside the subqueries, so there is no need for any filter conditions in the main query. One way to visualize what is going on is to run the subqueries and look at the result sets. Here are the results of the first subquery against the employee table:

```
mysql> SELECT emp_id, assigned_branch_id
    -> FROM employee
    -> WHERE start_date <= '2003-01-01'
    ->   AND (title = 'Teller' OR title = 'Head Teller');
+--------+--------------------+
| emp_id | assigned_branch_id |
+--------+--------------------+
|      8 |                  1 |
|      9 |                  1 |
|     10 |                  2 |
|     11 |                  2 |
```

```
|     13 |                   3 |
|     14 |                   3 |
|     16 |                   4 |
|     17 |                   4 |
|     18 |                   4 |
+--------+--------------------+
9 rows in set (0.03 sec)
```

Thus, this result set consists of a set of employee IDs and their corresponding branch IDs. When joined to the account table via the emp_id column, you now have an intermediate result set consisting of all rows from the account table with the additional column holding the branch ID of the employee that opened each account. Here are the results of the second subquery against the branch table:

```
mysql> SELECT branch_id
    -> FROM branch
    -> WHERE name = 'Woburn Branch';
+-----------+
| branch_id |
+-----------+
|         2 |
+-----------+
1 row in set (0.02 sec)
```

This query returns a single row containing a single column: the ID of the Woburn branch. This table is joined to the assigned_branch_id column of the intermediate result set, causing all accounts opened by non-Woburn-based employees to be filtered out of the final result set.

Using the Same Table Twice

If you are joining multiple tables, you might find that you need to join the same table more than once. In the sample database, for example, there are foreign keys to the branch table from both the account table (the branch at which the account was opened) and the employee table (the branch at which the employee works). If you want to include *both* branches in your result set, you can include the branch table twice in the from clause, joined once to the employee table and once to the account table. For this to work, you will need to give each instance of the branch table a different alias so the server knows which one you are referring to, as in:

```
mysql> SELECT a.account_id, e.emp_id,
    ->   b_a.name open_branch, b_e.name emp_branch
    -> FROM account a INNER JOIN branch b_a
    ->   ON a.open_branch_id = b_a.branch_id
    ->   INNER JOIN employee e
    ->   ON a.open_emp_id = e.emp_id
    ->   INNER JOIN branch b_e
    ->   ON e.assigned_branch_id = b_e.branch_id
    -> WHERE a.product_cd = 'CHK';
```

```
+------------+--------+----------------+----------------+
| account_id | emp_id | open_branch    | emp_branch     |
+------------+--------+----------------+----------------+
|          8 |      1 | Headquarters   | Headquarters   |
|         12 |      1 | Headquarters   | Headquarters   |
|         17 |      1 | Headquarters   | Headquarters   |
|          1 |     10 | Woburn Branch  | Woburn Branch  |
|          4 |     10 | Woburn Branch  | Woburn Branch  |
|          6 |     13 | Quincy Branch  | Quincy Branch  |
|         11 |     16 | So. NH Branch  | So. NH Branch  |
|         15 |     16 | So. NH Branch  | So. NH Branch  |
|         20 |     16 | So. NH Branch  | So. NH Branch  |
|         23 |     16 | So. NH Branch  | So. NH Branch  |
+------------+--------+----------------+----------------+
10 rows in set (0.07 sec)
```

This query shows who opened each checking account, what branch it was opened at, and to which branch the employee who opened the account is currently assigned. The branch table is included twice, with aliases b_a and b_e. By assigning different aliases to each instance of the branch table, the server is able to understand which instance you are referring to: the one joined to the account table, or the one joined to the employee table. Therefore, this is one example of a query that *requires* the use of table aliases.

Self-Joins

Not only can you include the same table more than once in the same query, you can actually join a table to itself. This might seem like a strange thing to do at first, but there are valid reasons for doing so. The employee table, for example, includes a *self-referencing foreign key*, which means that it includes a column (superior_emp_id) that points to the primary key within the same table. This column points to the employee's manager (unless the employee is the head honcho, in which case the column is null). Using a *self-join*, you can write a query that lists every employee's name along with the name of their manager:

```
mysql> SELECT e.fname, e.lname,
    ->   e_mgr.fname mgr_fname, e_mgr.lname mgr_lname
    -> FROM employee e INNER JOIN employee e_mgr
    ->   ON e.superior_emp_id = e_mgr.emp_id;
+----------+-----------+-----------+-----------+
| fname    | lname     | mgr_fname | mgr_lname |
+----------+-----------+-----------+-----------+
| Susan    | Barker    | Michael   | Smith     |
| Robert   | Tyler     | Michael   | Smith     |
| Susan    | Hawthorne | Robert    | Tyler     |
| John     | Gooding   | Susan     | Hawthorne |
| Helen    | Fleming   | Susan     | Hawthorne |
| Chris    | Tucker    | Helen     | Fleming   |
| Sarah    | Parker    | Helen     | Fleming   |
| Jane     | Grossman  | Helen     | Fleming   |
```

```
| Paula    | Roberts  | Susan    | Hawthorne |
| Thomas   | Ziegler  | Paula    | Roberts   |
| Samantha | Jameson  | Paula    | Roberts   |
| John     | Blake    | Susan    | Hawthorne |
| Cindy    | Mason    | John     | Blake     |
| Frank    | Portman  | John     | Blake     |
| Theresa  | Markham  | Susan    | Hawthorne |
| Beth     | Fowler   | Theresa  | Markham   |
| Rick     | Tulman   | Theresa  | Markham   |
+----------+----------+----------+-----------+
17 rows in set (0.01 sec)
```

This query includes two instances of the employee table, one to provide employee names (with the table alias e), and the other to provide manager names (with the table alias e_mgr). The on subclause uses these aliases to join the employee table to itself via the superior_emp_id foreign key. This is another example of a query for which table aliases are required; otherwise, the server wouldn't know whether you are referring to an employee or an employee's manager.

While there are 18 rows in the employee table, only 17 rows were returned by the query; the president of the bank, Michael Smith, has no superior (his superior_emp_id column is null), so the join failed for his row. To include Michael Smith in the result set, you would need to use an outer join, which will be covered in Chapter 10.

Equi-Joins Versus Non-Equi-Joins

All of the multitable queries shown thus far have employed *equi-joins*, meaning that values from the two tables must match for the join to succeed. An equi-join always employs an equals sign, as in:

```
ON e.assigned_branch_id = b.branch_id
```

While the majority of your queries will employ equi-joins, you can also join your tables via ranges of values, which are referred to as *non-equi-joins*. Here's an example of a query that joins by a range of values:

```
SELECT e.emp_id, e.fname, e.lname, e.start_date
FROM employee e INNER JOIN product p
  ON e.start_date >= p.date_offered
    AND e.start_date <= p.date_retired
WHERE p.name = 'no-fee checking';
```

This query joins two tables that have no foreign-key relationships. The intent is to find all employees who began working for the bank while the No-Fee Checking product was being offered. Thus, an employee's start date must be between the date the product was offered and the date the product was retired.

You may also find a need for a *self-non-equi-join*, meaning that a table is joined to itself using a non-equi-join. For example, let's say that the operations manager has decided to have a chess tournament for all bank tellers. You have been asked to create a list of all the pairings. You might try joining the employee table to itself for all tellers (title = 'Teller') and return all rows where the emp_id's don't match (since a person can't play chess with him/herself):

```
mysql> SELECT e1.fname, e1.lname, 'VS' vs, e2.fname, e2.lname
    -> FROM employee e1 INNER JOIN employee e2
    ->   ON e1.emp_id != e2.emp_id
    -> WHERE e1.title = 'Teller' AND e2.title = 'Teller';
+----------+----------+----+----------+----------+
| fname    | lname    | vs | fname    | lname    |
+----------+----------+----+----------+----------+
| Sarah    | Parker   | VS | Chris    | Tucker   |
| Jane     | Grossman | VS | Chris    | Tucker   |
| Thomas   | Ziegler  | VS | Chris    | Tucker   |
| Samantha | Jameson  | VS | Chris    | Tucker   |
| Cindy    | Mason    | VS | Chris    | Tucker   |
| Frank    | Portman  | VS | Chris    | Tucker   |
| Beth     | Fowler   | VS | Chris    | Tucker   |
| Rick     | Tulman   | VS | Chris    | Tucker   |
| Chris    | Tucker   | VS | Sarah    | Parker   |
| Jane     | Grossman | VS | Sarah    | Parker   |
| Thomas   | Ziegler  | VS | Sarah    | Parker   |
| Samantha | Jameson  | VS | Sarah    | Parker   |
| Cindy    | Mason    | VS | Sarah    | Parker   |
| Frank    | Portman  | VS | Sarah    | Parker   |
| Beth     | Fowler   | VS | Sarah    | Parker   |
| Rick     | Tulman   | VS | Sarah    | Parker   |
...
| Chris    | Tucker   | VS | Rick     | Tulman   |
| Sarah    | Parker   | VS | Rick     | Tulman   |
| Jane     | Grossman | VS | Rick     | Tulman   |
| Thomas   | Ziegler  | VS | Rick     | Tulman   |
| Samantha | Jameson  | VS | Rick     | Tulman   |
| Cindy    | Mason    | VS | Rick     | Tulman   |
| Frank    | Portman  | VS | Rick     | Tulman   |
| Beth     | Fowler   | VS | Rick     | Tulman   |
+----------+----------+----+----------+----------+
72 rows in set (0.01 sec)
```

You're on the right track, but the problem here is that for each pairing (i.e., Sarah Parker versus Chris Tucker), there is also a reverse pairing (i.e., Chris Tucker versus Sarah Parker). One way to achieve the desired results is to use the join condition e1.emp_id < e2.emp_id so that each teller is paired only with those tellers having a higher employee ID (you can also use e1.emp_id > e2.emp_id if you wish):

```
mysql> SELECT e1.fname, e1.lname, 'VS' vs, e2.fname, e2.lname
    -> FROM employee e1 INNER JOIN employee e2
    ->   ON e1.emp_id < e2.emp_id
    -> WHERE e1.title = 'Teller' AND e2.title = 'Teller';
+----------+----------+----+----------+----------+
| fname    | lname    | vs | fname    | lname    |
+----------+----------+----+----------+----------+
| Chris    | Tucker   | VS | Sarah    | Parker   |
```

```
| Chris    | Tucker    | VS | Jane     | Grossman |
| Chris    | Tucker    | VS | Thomas   | Ziegler  |
| Chris    | Tucker    | VS | Samantha | Jameson  |
| Chris    | Tucker    | VS | Cindy    | Mason    |
| Chris    | Tucker    | VS | Frank    | Portman  |
| Chris    | Tucker    | VS | Beth     | Fowler   |
| Chris    | Tucker    | VS | Rick     | Tulman   |
| Sarah    | Parker    | VS | Jane     | Grossman |
| Sarah    | Parker    | VS | Thomas   | Ziegler  |
| Sarah    | Parker    | VS | Samantha | Jameson  |
| Sarah    | Parker    | VS | Cindy    | Mason    |
| Sarah    | Parker    | VS | Frank    | Portman  |
| Sarah    | Parker    | VS | Beth     | Fowler   |
| Sarah    | Parker    | VS | Rick     | Tulman   |
| Jane     | Grossman  | VS | Thomas   | Ziegler  |
| Jane     | Grossman  | VS | Samantha | Jameson  |
| Jane     | Grossman  | VS | Cindy    | Mason    |
| Jane     | Grossman  | VS | Frank    | Portman  |
| Jane     | Grossman  | VS | Beth     | Fowler   |
| Jane     | Grossman  | VS | Rick     | Tulman   |
| Thomas   | Ziegler   | VS | Samantha | Jameson  |
| Thomas   | Ziegler   | VS | Cindy    | Mason    |
| Thomas   | Ziegler   | VS | Frank    | Portman  |
| Thomas   | Ziegler   | VS | Beth     | Fowler   |
| Thomas   | Ziegler   | VS | Rick     | Tulman   |
| Samantha | Jameson   | VS | Cindy    | Mason    |
| Samantha | Jameson   | VS | Frank    | Portman  |
| Samantha | Jameson   | VS | Beth     | Fowler   |
| Samantha | Jameson   | VS | Rick     | Tulman   |
| Cindy    | Mason     | VS | Frank    | Portman  |
| Cindy    | Mason     | VS | Beth     | Fowler   |
| Cindy    | Mason     | VS | Rick     | Tulman   |
| Frank    | Portman   | VS | Beth     | Fowler   |
| Frank    | Portman   | VS | Rick     | Tulman   |
| Beth     | Fowler    | VS | Rick     | Tulman   |
+----------+-----------+----+----------+----------+
36 rows in set (0.01 sec)
```

You now have a list of 36 pairings, which is the correct number when choosing pairs of nine distinct things.

Join Conditions Versus Filter Conditions

You are now familiar with the concept that join conditions belong in the on sub-clause, while filter conditions belong in the where clause. However, SQL is flexible as to where you place your conditions, so you will need to take care when constructing your queries. For example, the following query joins two tables using a single join clause, and also includes a single filter condition in the where clause:

```
mysql> SELECT a.account_id, a.product_cd, c.fed_id
    -> FROM account a INNER JOIN customer c
    ->   ON a.cust_id = c.cust_id
```

```
    -> WHERE c.cust_type_cd = 'B';
+------------+------------+------------+
| account_id | product_cd | fed_id     |
+------------+------------+------------+
|         20 | CHK        | 04-1111111 |
|         21 | BUS        | 04-1111111 |
|         22 | BUS        | 04-2222222 |
|         23 | CHK        | 04-3333333 |
|         24 | SBL        | 04-4444444 |
+------------+------------+------------+
5 rows in set (0.08 sec)
```

Pretty straightforward, but what happens if you mistakenly put the filter condition in the on subclause instead of in the where clause?

```
mysql> SELECT a.account_id, a.product_cd, c.fed_id
    -> FROM account a INNER JOIN customer c
    ->   ON a.cust_id = c.cust_id
    ->      AND c.cust_type_cd = 'B';
+------------+------------+------------+
| account_id | product_cd | fed_id     |
+------------+------------+------------+
|         20 | CHK        | 04-1111111 |
|         21 | BUS        | 04-1111111 |
|         22 | BUS        | 04-2222222 |
|         23 | CHK        | 04-3333333 |
|         24 | SBL        | 04-4444444 |
+------------+------------+------------+
5 rows in set (0.00 sec)
```

As you can see, the results of the second version, which has *both* conditions in the on subclause and has no where clause, generates the same results. What if both conditions are placed in the where clause but the from clause still uses the ANSI join syntax?

```
mysql> SELECT a.account_id, a.product_cd, c.fed_id
    -> FROM account a INNER JOIN customer c
    -> WHERE a.cust_id = c.cust_id
    ->   AND c.cust_type_cd = 'B';
+------------+------------+------------+
| account_id | product_cd | fed_id     |
+------------+------------+------------+
|         20 | CHK        | 04-1111111 |
|         21 | BUS        | 04-1111111 |
|         22 | BUS        | 04-2222222 |
|         23 | CHK        | 04-3333333 |
|         24 | SBL        | 04-4444444 |
+------------+------------+------------+
5 rows in set (0.00 sec)
```

Once again, the MySQL server has generated the same result set. It will be up to you to put your conditions in the proper place so that your queries are easy to understand and maintain.

Exercises

The following exercises are designed to test your understanding of inner joins. Please see Appendix C for solutions to these exercises.

5-1

Fill in the blanks (denoted by `<#>`) for the following query to obtain the results that follow:

```
mysql> SELECT e.emp_id, e.fname, e.lname, b.name
    -> FROM employee e INNER JOIN <1> b
    ->   ON e.assigned_branch_id = b.<2>;
+--------+----------+-----------+---------------+
| emp_id | fname    | lname     | name          |
+--------+----------+-----------+---------------+
|      1 | Michael  | Smith     | Headquarters  |
|      2 | Susan    | Barker    | Headquarters  |
|      3 | Robert   | Tyler     | Headquarters  |
|      4 | Susan    | Hawthorne | Headquarters  |
|      5 | John     | Gooding   | Headquarters  |
|      6 | Helen    | Fleming   | Headquarters  |
|      7 | Chris    | Tucker    | Headquarters  |
|      8 | Sarah    | Parker    | Headquarters  |
|      9 | Jane     | Grossman  | Headquarters  |
|     10 | Paula    | Roberts   | Woburn Branch |
|     11 | Thomas   | Ziegler   | Woburn Branch |
|     12 | Samantha | Jameson   | Woburn Branch |
|     13 | John     | Blake     | Quincy Branch |
|     14 | Cindy    | Mason     | Quincy Branch |
|     15 | Frank    | Portman   | Quincy Branch |
|     16 | Theresa  | Markham   | So. NH Branch |
|     17 | Beth     | Fowler    | So. NH Branch |
|     18 | Rick     | Tulman    | So. NH Branch |
+--------+----------+-----------+---------------+
18 rows in set (0.03 sec)
```

5-2

Write a query that returns the account ID for each nonbusiness customer (customer. cust_type_cd = 'I') along with the customer's federal ID (customer.fed_id) and the name of the product on which the account is based (product.name).

5-3

Construct a query that finds all employees whose supervisor is assigned to a different department. Retrieve the employees' ID, first name, and last name.

CHAPTER 6

Working with Sets

Although you can interact with the data in a database one row at a time, relational databases are really all about sets. You have seen how tables can be created via queries or subqueries, made persistent via `insert` statements, and brought together via joins; this chapter will explore how multiple tables can be combined using various set operators.

Set Theory Primer

In many parts of the world, basic set theory is included in elementary-level math curriculums. Perhaps you recall looking at something like what is shown in Figure 6-1.

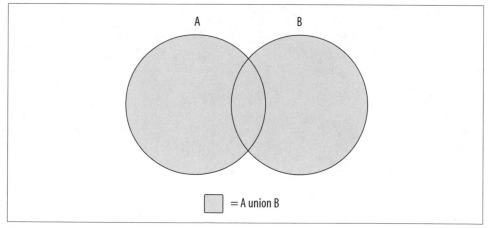

Figure 6-1. The union operation

The shaded area in Figure 6-1 represents the *union* of sets A and B, which is the combination of the two sets (with any overlapping regions included only once). Is this starting to look familiar? If so, then you'll finally get a chance to put that knowledge to use; if not, don't worry, because it's easy to visualize using a couple of diagrams.

Using circles to represent two data sets (A and B), imagine a subset of data that is common to both sets; this common data is represented by the overlapping area shown in Figure 6-1. Since set theory is rather uninteresting without an overlap between data sets, I will use the same diagram to illustrate each set operation. There is another set operation that is concerned *only* with the overlap between two data sets; this operation is known as the *intersection* and is demonstrated in Figure 6-2.

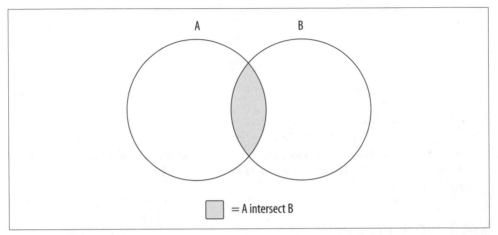

Figure 6-2. *The intersect operation*

The data set generated by the intersection of sets A and B is just the area of overlap between the two sets. If the two sets have no overlap, then the intersection operation yields the empty set.

The third and final set operation, which is demonstrated in Figure 6-3, is known as the *except* operation.

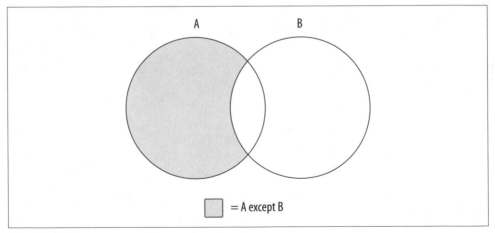

Figure 6-3. *The except operation*

Figure 6-3 shows the results of A except B, which is the whole of set A minus any overlap with set B. If the two sets have no overlap, then the operation A except B yields the whole of set A.

Using these three operations, or by combining different operations together, you can generate whatever results you need. For example, imagine that you want to build a set demonstrated by Figure 6-4.

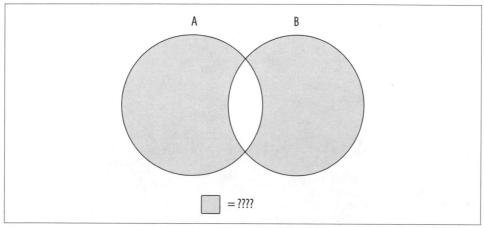

Figure 6-4. Mystery data set

The data set you are looking for includes all of sets A and B *without* the overlapping region. You can't achieve this outcome with just one of the three operations shown earlier; instead, you will need to first build a data set that encompasses all of sets A and B, and then utilize a second operation to remove the overlapping region. If the combined set is described as A union B, and the overlapping region is described as A intersect B, then the operation needed to generate the data set represented by Figure 6-4 would look as follows:

    (A union B) except (A intersect B)

Of course, there are often multiple ways to achieve the same results; a similar outcome could be reached using the following operation:

    (A except B) union (B except A)

While these concepts are fairly easy to understand using diagrams, the next sections will show you how these concepts are applied to a relational database using the SQL set operators.

Set Theory in Practice

The circles used in the previous section's diagrams to represent data sets don't convey anything about what the data sets comprise. When dealing with actual data, however, there is a need to describe the composition of the tables involved if they are to be combined. Imagine, for example, what would happen if you tried to generate the union of the product table and the customer table, whose table definitions are shown below:

```
mysql> DESC product;
+-----------------+-------------+------+-----+---------+-------+
| Field           | Type        | Null | Key | Default | Extra |
+-----------------+-------------+------+-----+---------+-------+
| product_cd      | varchar(10) |      | PRI |         |       |
| name            | varchar(50) |      |     |         |       |
| product_type_cd | varchar(10) |      | MUL |         |       |
| date_offered    | date        | YES  |     | NULL    |       |
| date_retired    | date        | YES  |     | NULL    |       |
+-----------------+-------------+------+-----+---------+-------+
5 rows in set (0.23 sec)

mysql> DESC customer;
+--------------+------------------+------+-----+---------+----------------+
| Field        | Type             | Null | Key | Default | Extra          |
+--------------+------------------+------+-----+---------+----------------+
| cust_id      | int(10) unsigned |      | PRI | NULL    | auto_increment |
| fed_id       | varchar(12)      |      |     |         |                |
| cust_type_cd | enum('I','B')    |      |     | I       |                |
| address      | varchar(30)      | YES  |     | NULL    |                |
| city         | varchar(20)      | YES  |     | NULL    |                |
| state        | varchar(20)      | YES  |     | NULL    |                |
| postal_code  | varchar(10)      | YES  |     | NULL    |                |
+--------------+------------------+------+-----+---------+----------------+
7 rows in set (0.04 sec)
```

When combined, the first column in the resulting table would be the combination of the product.product_cd and customer.cust_id columns, the second column would be the combination of the product.name and customer.fed_id columns, etc. While some of the column pairs are easy to combine (i.e., two numeric columns), it is unclear how other column pairs should be combined, such as a numeric column with a string column or a string column with a date column. Additionally, the sixth and seventh columns of the combined tables would only include data from the customer table's sixth and seventh columns, since the product table has only five columns. Clearly, there needs to be some commonality between two tables that you wish to combine.

Therefore, when performing set operations on actual tables, the following guidelines must apply:

- Both tables must have the same number of columns.
- The data types of each column across the two tables must be the same (or the server must be able to convert one to the other).

With these rules in place, it is easier to envision what "overlapping data" means in practice; each column pair from the two tables being combined must contain the same string, number, or date for rows in the two tables to be considered the same.

Set operations are performed by placing a *set operator* between two select statements, as demonstrated by the following:

```
mysql> SELECT 1 num, 'abc' str
    -> UNION
    -> SELECT 9 num, 'xyz' str;
+-----+-----+
| num | str |
+-----+-----+
|   1 | abc |
|   9 | xyz |
+-----+-----+
2 rows in set (0.02 sec)
```

Each of the individual queries yields a table consisting of a single row having a numeric column and a string column. The set operator, which, in this case, is union, tells the database server to combine all rows from the two tables. Thus, the final table includes two rows of two columns. This query is known as a *compound query* because it comprises multiple, otherwise-independent queries. As you will see later, compound queries may include *more* than two queries if multiple set operations are needed to attain the final results.

Set Operators

The SQL language includes three set operators that allow you to perform each of the various set operations described earlier in the chapter. Additionally, each set operator has two flavors, one that includes duplicates and another that removes duplicates (but not necessarily *all* of the duplicates). The following subsections will define each operator and demonstrate how they are used.

The union Operator

The union and union all operators allow you to combine multiple tables. The difference is that, when you want to combine two tables including *all* rows from both tables in the final result, even at the expense of having duplicates, you need to use the union all operator. With union all, the number of rows in the final table will always equal the sum of the number of rows in the original tables. This operation is the simplest set operation to perform (for the server's point of view), since there is no need for the server to check for overlapping data. The following example demonstrates how the union all operator can be used to generate a full set of customer data from the two customer subtype tables:

```
mysql> SELECT cust_id, lname name
    -> FROM individual
```

```
    -> UNION ALL
    -> SELECT cust_id, name
    -> FROM business;
+---------+-----------------------+
| cust_id | name                  |
+---------+-----------------------+
|       1 | Hadley                |
|       2 | Tingley               |
|       3 | Tucker                |
|       4 | Hayward               |
|       5 | Frasier               |
|       6 | Spencer               |
|       7 | Young                 |
|       8 | Blake                 |
|       9 | Farley                |
|      10 | Chilton Engineering   |
|      11 | Northeast Cooling Inc.|
|      12 | Superior Auto Body    |
|      13 | AAA Insurance Inc.    |
+---------+-----------------------+
13 rows in set (0.04 sec)
```

The query returns all 13 customers, with 9 rows coming from the individual table and the other 4 coming from the business table. While the business table includes a single column to hold the company name, the individual table includes two name columns, one each for the person's first and last names. In this case, I chose to include only the last name from the individual table.

Just to drive home the point that the union all operator doesn't remove duplicates, here's the same query as the previous example but with an additional query against the business table:

```
mysql> SELECT cust_id, lname name
    -> FROM individual
    -> UNION ALL
    -> SELECT cust_id, name
    -> FROM business
    -> UNION ALL
    -> SELECT cust_id, name
    -> FROM business;
+---------+-----------------------+
| cust_id | name                  |
+---------+-----------------------+
|       1 | Hadley                |
|       2 | Tingley               |
|       3 | Tucker                |
|       4 | Hayward               |
|       5 | Frasier               |
|       6 | Spencer               |
|       7 | Young                 |
|       8 | Blake                 |
|       9 | Farley                |
|      10 | Chilton Engineering   |
```

```
|     11 | Northeast Cooling Inc. |
|     12 | Superior Auto Body     |
|     13 | AAA Insurance Inc.     |
|     10 | Chilton Engineering    |
|     11 | Northeast Cooling Inc. |
|     12 | Superior Auto Body     |
|     13 | AAA Insurance Inc.     |
+---------+-----------------------+
17 rows in set (0.01 sec)
```

This compound query includes three select statements, two of which are identical.
As you can see by the results, the four rows from the business table are included
twice (customer IDs 10, 11, 12, and 13).

While you are unlikely to repeat the same query twice in a compound query, here is
another compound query that returns duplicate data:

```
mysql> SELECT emp_id
    -> FROM employee
    -> WHERE assigned_branch_id = 2
    ->   AND (title = 'Teller' OR title = 'Head Teller')
    -> UNION ALL
    -> SELECT DISTINCT open_emp_id
    -> FROM account
    -> WHERE open_branch_id = 2;
+--------+
| emp_id |
+--------+
|     10 |
|     11 |
|     12 |
|     10 |
+--------+
4 rows in set (0.01 sec)
```

The first query in the compound statement retrieves all tellers assigned to the
Woburn branch, whereas the second query returns the distinct set of tellers who
opened accounts at the Woburn branch. Of the four rows in the result set, one of
them is a duplicate (employee ID 10). If you would like your combined table to
exclude duplicate rows, you need to use the union operator instead of union all:

```
mysql> SELECT emp_id
    -> FROM employee
    -> WHERE assigned_branch_id = 2
    ->   AND (title = 'Teller' OR title = 'Head Teller')
    -> UNION
    -> SELECT DISTINCT open_emp_id
    -> FROM account
    -> WHERE open_branch_id = 2;
+--------+
| emp_id |
+--------+
|     10 |
|     11 |
```

```
|      12 |
+---------+
3 rows in set (0.01 sec)
```

For this version of the query, only the three distinct rows are included in the result set, rather than the four rows (three distinct, one duplicate) returned when using union all.

The intersect Operator

The ANSI SQL specification includes the intersect operator for performing intersections. Unfortunately, Version 4.1 of MySQL does not implement the intersect operator. If you are using Oracle (but not SQL Server), you will be able to use intersect; since I am using MySQL for all examples in this book, however, the result sets for the example queries in this section are fabricated and cannot be executed with any versions up to and including Version 5.0. I will also refrain from showing the MySQL prompt (mysql>), since the statements are not being executed by the MySQL server.

If the two queries in a compound query return nonoverlapping tables, then the intersection will be an empty set. Consider the following query:

```
SELECT emp_id, fname, lname
FROM employee
INTERSECT
SELECT cust_id, fname, lname
FROM individual;
Empty set (0.04 sec)
```

The first query returns the ID and name of each employee, while the second query returns the ID and name of each customer. These sets are completely nonoverlapping, so the intersection of the two sets yields the empty set.

The next step is to identify two queries that *do* have overlapping data and then apply the intersect operator. For this purpose, I will use the same query used to demonstrate the difference between union and union all, except this time using intersect:

```
SELECT emp_id
FROM employee
WHERE assigned_branch_id = 2
  AND (title = 'Teller' OR title = 'Head Teller')
INTERSECT
SELECT DISTINCT open_emp_id
FROM account
WHERE open_branch_id = 2;
+---------+
| emp_id |
+---------+
|      10 |
+---------+
1 row in set (0.01 sec)
```

The intersection of these two queries yields employee ID 10, which is the only value found in both queries' result sets.

Along with the intersect operator, which removes any duplicate rows found in the overlapping region, the ANSI SQL specification calls for an intersect all operator, which does not remove duplicates. The only database server that currently implements the intersect all operator is IBM's DB2 Universal Server.

The except Operator

The ANSI SQL specification includes the except operator for performing the except operation. Once again, unfortunately, Version 4.1 of MySQL does not implement the except operator, so the same rules apply for this section as for the previous section.

 If you are using Oracle Database, you will need to use the non-ANSI-compliant minus operator instead.

The except operation returns the first table minus any overlap with the second table. Here's the example from the previous section, but using except instead of intersect:

```
SELECT emp_id
FROM employee
WHERE assigned_branch_id = 2
  AND (title = 'Teller' OR title = 'Head Teller')
EXCEPT
SELECT DISTINCT open_emp_id
FROM account
WHERE open_branch_id = 2;
+--------+
| emp_id |
+--------+
|     11 |
|     12 |
+--------+
2 rows in set (0.01 sec)
```

In this version of the query, the result set consists of the three rows from the first query minus employee ID 10, which is found in the result sets from both queries. There is also an except all operator specified in the ANSI SQL specification, but, once again, only IBM's DB2 Universal Server has implemented the except all operator.

The except all operator is a bit tricky, so here's an example to demonstrate how duplicate data is handled. Let's say you have two data sets that look as follows:

Set A

```
+--------+
| emp_id |
+--------+
|     10 |
|     11 |
|     12 |
|     10 |
|     10 |
+--------+
```

Set B

```
+--------+
| emp_id |
+--------+
|     10 |
|     10 |
+--------+
```

The operation A except B yields the following:

```
+--------+
| emp_id |
+--------+
|     11 |
|     12 |
+--------+
```

If you change the operation to A except all B, you will see the following:

```
+--------+
| emp_id |
+--------+
|     10 |
|     11 |
|     12 |
+--------+
```

Therefore, the difference between the two operations is that except removes all occurrences of duplicate data from set A whereas except all only removes one occurrence of duplicate data from set A for every occurrence in set B.

Set Operation Rules

The following sections outline some rules that must be followed when working with compound queries.

Sorting Compound Query Results

If you want the results of your compound query to be sorted, you can add an order by clause after the last query. When specifying column names in the order by clause, you will need to choose from the column names in the first query of the compound query. In every example thus far in the chapter, the column names have been the same for both queries in the compound query, but this does not need to be the case, as demonstrated by the following:

```
mysql> SELECT emp_id, assigned_branch_id
    -> FROM employee
    -> WHERE title = 'Teller'
    -> UNION
    -> SELECT open_emp_id, open_branch_id
    -> FROM account
    -> WHERE product_cd = 'SAV'
    -> ORDER BY emp_id;
+--------+--------------------+
| emp_id | assigned_branch_id |
+--------+--------------------+
|      1 |                  1 |
|      7 |                  1 |
|      8 |                  1 |
|      9 |                  1 |
|     10 |                  2 |
|     11 |                  2 |
|     12 |                  2 |
|     14 |                  3 |
|     15 |                  3 |
|     16 |                  4 |
|     17 |                  4 |
|     18 |                  4 |
+--------+--------------------+
12 rows in set (0.04 sec)
```

The column names specified in the two queries are different in this example. If you specify a column name from the second query in your order by clause, you will see the following error:

```
mysql> SELECT emp_id, assigned_branch_id
    -> FROM employee
    -> WHERE title = 'Teller'
    -> UNION
    -> SELECT open_emp_id, open_branch_id
    -> FROM account
    -> WHERE product_cd = 'SAV'
    -> ORDER BY open_emp_id;
ERROR 1054 (42S22): Unknown column 'open_emp_id' in 'order clause'
```

I recommend giving the columns in both queries identical column aliases in order to avoid this issue.

Set Operation Precedence

If your compound query contains more than two queries using different set operators, you need to think about the order in which to place the queries in your compound statement to achieve the desired results. Consider the following three-query compound statement:

```
mysql> SELECT cust_id
    -> FROM account
    -> WHERE product_cd IN ('SAV', 'MM')
    -> UNION ALL
    -> SELECT a.cust_id
    -> FROM account a INNER JOIN branch b
    ->   ON a.open_branch_id = b.branch_id
    -> WHERE b.name = 'Woburn Branch'
    -> UNION
    -> SELECT cust_id
    -> FROM account
    -> WHERE avail_balance BETWEEN 500 AND 2500;
+---------+
| cust_id |
+---------+
|       1 |
|       2 |
|       3 |
|       4 |
|       8 |
|       9 |
|       7 |
|      11 |
|       5 |
+---------+
9 rows in set (0.00 sec)
```

This compound query includes three queries that return sets of nonunique customer IDs; the first two queries are separated with the union all operator, while the second and third queries are separated by the union operator. While it might not seem to make much difference where the union and union all operators are placed, it does, in fact, make a difference. Here's the same compound query with the set operators reversed:

```
mysql> SELECT cust_id
    -> FROM account
    -> WHERE product_cd IN ('SAV', 'MM')
    -> UNION
    -> SELECT a.cust_id
    -> FROM account a INNER JOIN branch b
    ->   ON a.open_branch_id = b.branch_id
    -> WHERE b.name = 'Woburn Branch'
    -> UNION ALL
    -> SELECT cust_id
    -> FROM account
    -> WHERE avail_balance BETWEEN 500 AND 2500;
```

```
+---------+
| cust_id |
+---------+
|       1 |
|       2 |
|       3 |
|       4 |
|       8 |
|       9 |
|       7 |
|      11 |
|       1 |
|       1 |
|       2 |
|       3 |
|       3 |
|       4 |
|       4 |
|       5 |
|       9 |
+---------+
17 rows in set (0.00 sec)
```

Looking at the results, it's obvious that it *does* make a difference how the compound query is arranged when using different set operators. In general, compound queries containing three or more queries are evaluated in order from top to bottom, but with the following caveats:

- The ANSI SQL specification calls for the intersect operator to have precedence over the other set operators.

- You may dictate the order in which queries are combined by enclosing multiple queries in parentheses.

However, since MySQL does not yet implement intersect nor allow parentheses in compound queries, you will need to carefully arrange the queries in your compound query so that you achieve the desired results. If you are using a different database server, you can wrap adjoining queries in parentheses to override the default top-to-bottom processing of compound queries, as in:

```
(SELECT cust_id
 FROM account
 WHERE product_cd IN ('SAV', 'MM')
 UNION ALL
 SELECT a.cust_id
 FROM account a INNER JOIN branch b
   ON a.open_branch_id = b.branch_id
 WHERE b.name = 'Woburn Branch')
INTERSECT
(SELECT cust_id
 FROM account
 WHERE avail_balance BETWEEN 500 AND 2500
 EXCEPT
```

```
SELECT cust_id
FROM account
WHERE product_cd = 'CD'
  AND avail_balance < 1000);
```

For this compound query, the first and second queries would be combined using the union all operator, then the third and fourth queries would be combined using the except operator, and, finally, the results from these two operations would be combined using the intersect operator to generate the final result set.

Exercises

The following exercises are designed to test your understanding of set operations. See Appendix C for answers to these exercises.

6-1

If set A = {L M N O P} and set B = {P Q R S T}, what sets are generated by the following operations:

- A union B
- A union all B
- A intersect B
- A except B

6-2

Write a compound query that finds the first and last names of all individual customers along with the first and last names of all employees.

6-3

Sort the results from exercise 6-2 by the lname column.

CHAPTER 7

Data Generation, Conversion, and Manipulation

As I mentioned in the Preface, this book strives to teach generic SQL techniques that can be applied across multiple database servers. This chapter, however, deals with the generation, conversion, and manipulation of string, numeric, and temporal data, and the SQL language does not include commands covering this functionality. Rather, built-in functions are used to facilitate data generation, conversion, and manipulation, and, while the SQL standard does specify some functions, the database vendors often do not comply with the function specifications.

Therefore, my approach for this chapter is to show you some of the common ways in which data is manipulated within SQL statements, and then demonstrate some of the built-in functions implemented by Microsoft SQL Server, Oracle Database, and MySQL. Along with this chapter, I strongly recommend you purchase a reference guide covering all of the functions implemented by your server. If you work with more than one database server, there are several reference guides that cover multiple servers, such as *SQL in a Nutshell* or *SQL Pocket Guide*, both from O'Reilly.

Working with String Data

When working with string data, you will be using one of the following character data types:

CHAR

> Holds fixed-length, blank-padded strings. MySQL allows CHAR values up to 255 characters in length, Oracle Database permits up to 2,000 characters, and SQL Server allows up to 8,000 characters.

varchar

> Holds variable-length strings. MySQL permits up to 255 characters in a varchar column (65,535 characters for Version 5.0 and up), Oracle Database (via the varchar2 type) allows up to 4,000 characters, and SQL Server allows up to 8,000 characters.

text *(MySQL and SQL Server)* or `CLOB` *(Character Large Object; Oracle Database)*

> Holds very large variable-length strings (generally referred to as documents in this context). MySQL has multiple text types (`tinytext`, `text`, `mediumtext`, and `longtext`) for documents up to 4 GB in size. SQL Server has a single text type for documents up to 2 GB in size, and Oracle Database includes the `CLOB` data type, which can hold documents up to a whopping 128 terabytes.

To demonstrate how these various types may be used, I will be using the following table for some of the examples in this section:

```
CREATE TABLE string_tbl
 (char_fld CHAR(30),
  vchar_fld VARCHAR(30),
  text_fld TEXT
 );
```

The next two subsections will show how string data may be generated and manipulated.

String Generation

The simplest way to populate a character column is to enclose a string in quotes, as in:

```
mysql> INSERT INTO string_tbl (char_fld, vchar_fld, text_fld)
    -> VALUES ('This is char data',
    ->    'This is varchar data',
    ->    'This is text data');
Query OK, 1 row affected (0.00 sec)
```

When inserting string data into a table, remember that if the length of the string exceeds the maximum size for the character column (either the designated maximum or the maximum allowed for the data type), the server will either throw an exception (Oracle Database), or, in the case of MySQL or SQL Server, quietly truncate the string (MySQL also generates a warning). To demonstrate how MySQL handles this situation, the following update statement attempts to modify the vchar_fld column, whose maximum length is defined as 30, with a string that is 46 characters in length:

```
mysql> UPDATE string_tbl
    -> SET vchar_fld = 'This is a piece of extremely long varchar data';
Query OK, 1 row affected, 1 warning (0.01 sec)
Rows matched: 1  Changed: 1  Warnings: 1
```

The column is modified, but the following warning is generated:

```
mysql> SHOW WARNINGS;
+---------+------+------------------------------------------------+
| Level   | Code | Message                                        |
+---------+------+------------------------------------------------+
| Warning | 1265 | Data truncated for column 'vchar_fld' at row 1 |
+---------+------+------------------------------------------------+
1 row in set (0.00 sec)
```

If you retrieve the vchar_fld column, you will see the following:

```
mysql> SELECT vchar_fld
    -> FROM string_tbl;
+-------------------------------+
| vchar_fld                     |
+-------------------------------+
| This is a piece of extremely l |
+-------------------------------+
1 row in set (0.05 sec)
```

As you can see, only the first 30 characters of the 46-character string made it into the vchar_fld column. The best way to avoid string truncation (or exceptions, in the case of Oracle Database) when working with varchar columns is to set the upper limit of a column to a high enough value to handle the longest strings that might be stored in the column (keeping in mind that the server allocates only enough space to store the string, so it is not wasteful to set a high upper limit for varchar columns).

Including single quotes

Since strings are demarcated by single quotes, you will need to be alert for strings that include single quotes or apostrophes. For example, you won't be able to insert the following string because the server will think that the apostrophe in the word "doesn't" marks the end of the string:

```
UPDATE string_tbl
SET text_fld = 'This string doesn't work';
```

To make the server ignore the apostrophe in the word "doesn't," you will need to add an *escape* to the string so that the server treats the apostrophe like any other character in the string. All three servers allow you to escape a single quote by adding another single quote directly before the apostrophe, as in:

```
mysql> UPDATE string_tbl
    -> SET text_fld = 'This string didn''t work, but it does now';
Query OK, 1 row affected (0.01 sec)
Rows matched: 1  Changed: 1  Warnings: 0
```

 Oracle Database and MySQL users may also choose to escape a single quote by adding a backslash character immediately before, as in:

```
UPDATE string_tbl SET text_fld =
    'This string didn\'t work, but it does now'
```

If you retrieve a string for use in a screen or report field, you don't need to do anything special to handle embedded quotes:

```
mysql> SELECT text_fld
    -> FROM string_tbl;
+-------------------------------------------+
| text_fld                                  |
+-------------------------------------------+
| This string didn't work, but it does now |
```

```
+---------------------------------------------+
1 row in set (0.00 sec)
```

However, if you are retrieving the string to add to a file that will be read by another program, you may want to include the escape as part of the retrieved string. If you are using MySQL, you can use the built-in function quote( ), which places quotes around the entire string *and* adds escapes to any single quotes/apostrophes within the string. Here's what our string looks like when retrieved via the quote( ) function:

```
mysql> SELECT QUOTE(text_fld)
    -> FROM string_tbl;
+-------------------------------------------+
| QUOTE(text_fld)                           |
+-------------------------------------------+
| 'This string didn\'t work, but it does now' |
+-------------------------------------------+
1 row in set (0.04 sec)
```

When retrieving data for data export, you may want to use the quote( ) function for all nonsystem-generated character columns, such as a customer_notes column.

Including special characters

If your application is multinational in scope, you might find yourself working with strings that include characters that do not appear on your keyboard. When working with the French or German languages, for example, you might need to include accented characters such as é or ö. The SQL Server and MySQL servers include the built-in function char( ) so that you can build strings from any of the 255 characters in the ASCII character set (Oracle Database users can use the chr( ) function). To demonstrate, the next example retrieves a typed string and its equivalent built via individual characters:

```
mysql> SELECT 'abcdefg', CHAR(97,98,99,100,101,102,103);
+---------+--------------------------------+
| abcdefg | CHAR(97,98,99,100,101,102,103) |
+---------+--------------------------------+
| abcdefg | abcdefg                        |
+---------+--------------------------------+
1 row in set (0.01 sec)
```

Thus, the 97th character in the ASCII character set is the letter a. While the characters shown in the preceding are not special, the following examples show the location of the accented characters along with other special characters, such as currency symbols:

```
mysql> SELECT CHAR(128,129,130,131,132,133,134,135,136,137);
+-----------------------------------------------+
| CHAR(128,129,130,131,132,133,134,135,136,137) |
+-----------------------------------------------+
| Çüéâäàåçêë                                     |
+-----------------------------------------------+
1 row in set (0.01 sec)
```

```
mysql> SELECT CHAR(138,139,140,141,142,143,144,145,146,147);
+-----------------------------------------------+
| CHAR(138,139,140,141,142,143,144,145,146,147) |
+-----------------------------------------------+
| èïîìÄÅÉæÆô                                     |
+-----------------------------------------------+
1 row in set (0.01 sec)

mysql> SELECT CHAR(148,149,150,151,152,153,154,155,156,157);
+-----------------------------------------------+
| CHAR(148,149,150,151,152,153,154,155,156,157) |
+-----------------------------------------------+
| öòûùÿ...Ü¢£¥                                   |
+-----------------------------------------------+
1 row in set (0.00 sec)

mysql> SELECT CHAR(158,159,160,161,162,163,164,165);
+---------------------------------------+
| CHAR(158,159,160,161,162,163,164,165) |
+---------------------------------------+
| ℞ƒáíóúñÑ                               |
+---------------------------------------+
1 row in set (0.01 sec)
```

> I am using the latin1 character set for the examples in this section. If
> your session is configured for a different character set, you will see a
> different set of characters than what is shown here. The same con-
> cepts apply, but you will need to familiarize yourself with the layout of
> your character set to locate specific characters.

Building strings character by character can be quite tedious, especially if only a few
of the characters in the string are accented. Fortunately, you can use the concat()
function to concatenate individual strings, some of which can be typed while others
are generated via the char() function. For example, the following shows how to
build the phrase *danke schön* using the concat() and char() functions:

```
mysql> SELECT CONCAT('danke sch', CHAR(148), 'n');
+-------------------------------------+
| CONCAT('danke sch', CHAR(148), 'n') |
+-------------------------------------+
| danke schön                         |
+-------------------------------------+
1 row in set (0.00 sec)
```

Oracle Database users can use the concatenation operator (||) instead of the concat( ) function, as in:

```
SELECT 'danke sch' || CHR(148) || 'n'
FROM dual;
```

SQL Server does not include a concat( ) function, so you will need to use the concatenation operator (+), as in:

```
SELECT 'danke sch' + CHAR(148) + 'n'
```

If you have a character and need to find its ASCII equivalent, you can use the ascii( ) function, which takes the leftmost character in the string and returns a number:

```
mysql> SELECT ASCII('ö');
+------------+
| ASCII('ö') |
+------------+
|        148 |
+------------+
1 row in set (0.00 sec)
```

Using the char( ), ascii( ), and concat( ) functions (or concatenation operators), you should be able to work with any Roman language even if you are using a keyboard that does not include accented or special characters.

String Manipulation

Each database server includes many built-in functions for manipulating strings. This section will explore two types of string functions: those that return numbers, and those that return strings. Before I begin, however, I will reset the data in the string_tbl table to the following:

```
mysql> DELETE FROM string_tbl;
Query OK, 1 row affected (0.02 sec)

mysql> INSERT INTO string_tbl (char_fld, vchar_fld, text_fld)
    -> VALUES ('This string is 28 characters',
    ->    'This string is 28 characters',
    ->    'This string is 28 characters');
Query OK, 1 row affected (0.00 sec)
```

String functions that return numbers

Of the string functions that return numbers, one of the most commonly used is the length( ) function, which returns the number of characters in the string (SQL Server users will need to use the len( ) function). The following query applies the length( ) function to each column in the string_tbl table:

```
mysql> SELECT LENGTH(char_fld) char_length,
    ->    LENGTH(vchar_fld) varchar_length,
    ->    LENGTH(text_fld) text_length
    -> FROM string_tbl;
```

```
+-------------+----------------+-------------+
| char_length | varchar_length | text_length |
+-------------+----------------+-------------+
|          28 |             28 |          28 |
+-------------+----------------+-------------+
1 row in set (0.00 sec)
```

While the lengths of the varchar and text columns are as expected, you might have expected the length of the char column to be 30, since I told you that strings stored in char columns are right-padded with spaces. The MySQL server removes trailing spaces from char data when it is retrieved, however, so you will see the same results from all string functions regardless of the type of column in which the strings are stored.

Along with finding the length of a string, you might want to find the location of a substring within a string. For example, if you want to find the position at which the string "characters" appears in the vchar_fld column, you could use the position() function, as demonstrated by the following:

```
mysql> SELECT POSITION('characters' IN vchar_fld)
    -> FROM string_tbl;
+-------------------------------------+
| POSITION('characters' IN vchar_fld) |
+-------------------------------------+
|                                  19 |
+-------------------------------------+
1 row in set (0.12 sec)
```

If the substring cannot be found, the position() function returns 0.

 For those of you who program in languages such as C or C++, where the first element of an array is at position 0, remember when working with databases that the first character in a string is at position 1. A return value of 0 from position() indicates that the substring could not be found, not that the substring was found at the first position in the string.

If you want to start your search at something other than the first character of your target string, you will need to use the locate() function, which is similar to the position() function except that it allows an optional third parameter, which is used to define the search's start position. The locate() function is also proprietary, whereas the position() function is part of the SQL:2003 standard. Here's an example asking for the position of the string 'is' starting at the fifth character in the vchar_fld column:

```
mysql> SELECT LOCATE('is', vchar_fld, 5)
    -> FROM string_tbl;
+----------------------------+
| LOCATE('is', vchar_fld, 5) |
+----------------------------+
```

```
|           13 |
+--------------------------+
1 row in set (0.02 sec)
```

 Oracle Database does not include the position( ) or locate( ) functions, but it does include the instr( ) function, which mimics the
position( ) function when provided with two arguments and mimics
the locate( ) function when provided with three arguments. SQL
Server also doesn't include a position( ) or locate( ) function, but it
does include the charindx( ) function, which also accepts either two or
three arguments similar to Oracle's instr( ) function.

Another function that takes strings as arguments and returns numbers is the string
comparison function strcmp( ). Strcmp( ), which is implemented only by MySQL and
has no analogue in Oracle Database or SQL Server, takes two strings as arguments,
and returns one of the following:

-1   If the first string comes before the second string in sort order

0   If the strings are identical

1   If the first string comes after the second string in sort order

To illustrate how the function works, I will first show the sort order of five strings
using a query, and then show how the strings compare to one another using strcmp( ).
Here are the five strings that I will insert into the string_tbl table:

```
mysql> DELETE FROM string_tbl;
Query OK, 1 row affected (0.00 sec)

mysql> INSERT INTO string_tbl(vchar_fld) VALUES ('abcd');
Query OK, 1 row affected (0.03 sec)

mysql> INSERT INTO string_tbl(vchar_fld) VALUES ('xyz');
Query OK, 1 row affected (0.00 sec)

mysql> INSERT INTO string_tbl(vchar_fld) VALUES ('QRSTUV');
Query OK, 1 row affected (0.00 sec)

mysql> INSERT INTO string_tbl(vchar_fld) VALUES ('qrstuv');
Query OK, 1 row affected (0.00 sec)

mysql> INSERT INTO string_tbl(vchar_fld) VALUES ('12345');
Query OK, 1 row affected (0.00 sec)
```

Here are the five strings in their sort order:

```
mysql> SELECT vchar_fld
    -> FROM string_tbl
    -> ORDER BY vchar_fld;
+-----------+
| vchar_fld |
+-----------+
```

```
| 12345    |
| abcd     |
| QRSTUV   |
| qrstuv   |
| xyz      |
+----------+
5 rows in set (0.00 sec)
```

The next query makes six comparisons between the five different strings:

```
mysql> SELECT STRCMP('12345','12345') 12345_12345,
    ->     STRCMP('abcd','xyz') abcd_xyz,
    ->     STRCMP('abcd','QRSTUV') abcd_QRSTUV,
    ->     STRCMP('qrstuv','QRSTUV') qrstuv_QRSTUV,
    ->     STRCMP('12345','xyz') 12345_xyz,
    ->     STRCMP('xyz','qrstuv') xyz_qrstuv;
+-------------+----------+-------------+---------------+-----------+------------+
| 12345_12345 | abcd_xyz | abcd_QRSTUV | qrstuv_QRSTUV | 12345_xyz | xyz_qrstuv |
+-------------+----------+-------------+---------------+-----------+------------+
|           0 |       -1 |          -1 |             0 |        -1 |          1 |
+-------------+----------+-------------+---------------+-----------+------------+
1 row in set (0.00 sec)
```

The first comparison yields 0, which is to be expected since I compared a string to
itself. The fourth comparison also 0yields, which is a bit surprising, since the strings
are composed of the same letters, with one string all uppercase and the other all low-
ercase. The reason for this result is that MySQL's strcmp( ) function is case insensi-
tive, which is something to remember when using the function. The other four
comparisons yield either a -1 or a 1 depending on whether the first string comes
before or after the second string in sort order. For example, strcmp('abcd','xyz')
yields -1, since the string "abcd" comes before the string 'xyz'.

Along with the strcmp( ) function, MySQL also allows you to use the like and
regexp operators to compare strings in the select clause. Such comparisons will yield
1 (for true) or 0 (for false). Therefore, these operators allow you to build expressions that return a number, much like the functions described in this section. Here's
an example using like:

```
mysql> SELECT name, name LIKE '%ns' ends_in_ns
    -> FROM department;
+----------------+------------+
| name           | ends_in_ns |
+----------------+------------+
| Operations     |          1 |
| Loans          |          1 |
| Administration |          0 |
+----------------+------------+
3 rows in set (0.25 sec)
```

This example retrieves all of the department names, along with an expression that
returns 1 if the department name ends in "ns" or a 0 otherwise. If you want to perform

more complex pattern matches, you can use the regexp operator, as demonstrated by the following:

```
mysql> SELECT cust_id, cust_type_cd, fed_id,
    ->   fed_id REGEXP '.{3}-.{2}-.{4}' is_ss_no_format
    -> FROM customer;
+---------+--------------+-------------+-----------------+
| cust_id | cust_type_cd | fed_id      | is_ss_no_format |
+---------+--------------+-------------+-----------------+
|       1 | I            | 111-11-1111 |               1 |
|       2 | I            | 222-22-2222 |               1 |
|       3 | I            | 333-33-3333 |               1 |
|       4 | I            | 444-44-4444 |               1 |
|       5 | I            | 555-55-5555 |               1 |
|       6 | I            | 666-66-6666 |               1 |
|       7 | I            | 777-77-7777 |               1 |
|       8 | I            | 888-88-8888 |               1 |
|       9 | I            | 999-99-9999 |               1 |
|      10 | B            | 04-1111111  |               0 |
|      11 | B            | 04-2222222  |               0 |
|      12 | B            | 04-3333333  |               0 |
|      13 | B            | 04-4444444  |               0 |
+---------+--------------+-------------+-----------------+
13 rows in set (0.00 sec)
```

The fourth column of this query returns a 1 if the value stored in the fed_id column matches the format for a Social Security number.

 SQL Server and Oracle Database users can achieve similar results by building case expressions, which are described in detail in Chapter 11.

String functions that return strings

In some cases, you will need to modify existing strings, either by extracting part of the string or by adding additional text to the string. Every database server includes multiple functions to help with these tasks. Before I begin, I will once again reset the data in the string_tbl table:

```
mysql> DELETE FROM string_tbl;
Query OK, 5 rows affected (0.00 sec)

mysql> INSERT INTO string_tbl (text_fld)
    -> VALUES ('This string was 29 characters');
Query OK, 1 row affected (0.01 sec)
```

Earlier in the chapter, I demonstrated the use of the concat( ) function to help build words that include accented characters. The concat( ) function is useful in many other situations, including when you need to append additional characters to a stored string. For example, the following example modifies the string stored in the text_fld column by tacking an additional phrase on the end:

```
mysql> UPDATE string_tbl
    -> SET text_fld = CONCAT(text_fld, ', but now it is longer');
Query OK, 1 row affected (0.03 sec)
Rows matched: 1  Changed: 1  Warnings: 0
```

The contents of the text_fld column are now as follows:

```
mysql> SELECT text_fld
    -> FROM string_tbl;
+--------------------------------------------------+
| text_fld                                         |
+--------------------------------------------------+
| This string was 29 characters, but now it is longer |
+--------------------------------------------------+
1 row in set (0.00 sec)
```

Thus, like all functions that return a string, concat( ) may be used to replace the data stored in a character column.

Another common use for the concat( ) function is to build a string from individual pieces of data. For example, the following query generates a narrative string for each bank teller:

```
mysql> SELECT CONCAT(fname, ' ', lname, ' has been a ',
    ->    title, ' since ', start_date) emp_narrative
    -> FROM employee
    -> WHERE title = 'Teller' OR title = 'Head Teller';
+----------------------------------------------------------+
| emp_narrative                                            |
+----------------------------------------------------------+
| Helen Fleming has been a Head Teller since 2004-03-17    |
| Chris Tucker has been a Teller since 2004-09-15          |
| Sarah Parker has been a Teller since 2002-12-02          |
| Jane Grossman has been a Teller since 2002-05-03         |
| Paula Roberts has been a Head Teller since 2002-07-27    |
| Thomas Ziegler has been a Teller since 2000-10-23        |
| Samantha Jameson has been a Teller since 2003-01-08      |
| John Blake has been a Head Teller since 2000-05-11       |
| Cindy Mason has been a Teller since 2002-08-09           |
| Frank Portman has been a Teller since 2003-04-01         |
| Theresa Markham has been a Head Teller since 2001-03-15  |
| Beth Fowler has been a Teller since 2002-06-29           |
| Rick Tulman has been a Teller since 2002-12-12           |
+----------------------------------------------------------+
13 rows in set (0.12 sec)
```

The concat( ) function can handle any expression that returns a string, and will even convert numbers and dates to string format, as evidenced by the date column (start_date) used as an argument. While Oracle Database includes the concat( ) function, it will only accept two string arguments, so the previous query will not work on Oracle. Instead, you would need to use the concatenation operator (||) instead of using a function call, as in:

```
SELECT fname || ' ' || lname || ' has been a ' ||
    title || ' since ' || start_date emp_narrative
FROM employee
WHERE title = 'Teller' OR title = 'Head Teller';
```

SQL Server does not include a concat( ) function, so you would need to use the same approach as the previous query, except that you would use SQL Server's concatenation operator (+) instead of ||.

While concat( ) is useful for adding characters to the beginning or end of a string, you may also have a need to add or replace characters in the *middle* of a string. All three database servers provide functions for this purpose, but all of them are different, so I will demonstrate the MySQL function and then show the functions from the other two servers.

MySQL includes the insert( ) function, which takes four arguments: the original string, the position at which to start, the number of characters to replace, and the replacement string. Depending on the value of the third argument, the function may be used to either insert or replace characters in a string. With a value of zero for the third argument, the replacement string is inserted and any trailing characters are pushed to the right, as in:

```
mysql> SELECT INSERT('goodbye world', 9, 0, 'cruel ') string;
+---------------------+
| string              |
+---------------------+
| goodbye cruel world |
+---------------------+
1 row in set (0.00 sec)
```

In this example, all characters starting from position 9 are pushed to the right and the string 'cruel ' is inserted. If the third argument is greater than zero, then that number of characters is replaced with the replacement string, as in:

```
mysql> SELECT INSERT('goodbye world', 1, 7, 'hello') string;
+-------------+
| string      |
+-------------+
| hello world |
+-------------+
1 row in set (0.00 sec)
```

For this example, the first seven characters are replaced with the string 'hello' Oracle Database does not provide a single function with the flexibility of MySQL's insert( ) function, but Oracle does provide the replace( ) function, which is useful for replacing a substring with another substring. Here's the previous example reworked to use replace( ):

```
SELECT REPLACE('goodbye world', 'goodbye', 'hello')
FROM dual;
```

All instances of the string 'goodbye' will be replaced with the string 'hello', resulting in the string 'hello world'. The replace( ) function will replace *every* instance of the search string with the replacement string, so you need to be careful that you don't end up with more replacements than you anticipated.

SQL Server also includes a replace( ) function with the same functionality as Oracle's, but SQL Server also includes a function called stuff( ) with similar functionality to MySQL's insert( ) function. Here's an example:

```
SELECT STUFF('hello world', 1, 5, 'goodbye cruel')
```

When executed, five characters are removed starting at position 1, and then the string 'goodbye cruel' is inserted at the starting position, resulting in the string 'goodbye cruel world'.

Along with inserting characters into a string, you may have a need to *extract* a substring from a string. For this purpose, all three servers include the substring( ) function (although Oracle Database's version is called substr( )), which extracts a specified number of characters starting at a specified position. The following example extracts five characters from a string starting at the ninth position:

```
mysql> SELECT SUBSTRING('goodbye cruel world', 9, 5);
+----------------------------------------+
| SUBSTRING('goodbye cruel world', 9, 5) |
+----------------------------------------+
| cruel                                  |
+----------------------------------------+
1 row in set (0.00 sec)
```

Along with the functions demonstrated here, all three servers include many more built-in functions for manipulating string data. While many of them are designed for very specific purposes, such as generating the string equivalent of octal or hexadecimal numbers, there are many other general-purpose functions as well, such as functions that remove or add trailing spaces. For more information, consult your server's SQL reference guide or a general-purpose SQL reference guide such as *SQL in a Nutshell* (O'Reilly).

Working with Numeric Data

Unlike string data (and temporal data, as you will see shortly), numeric data generation is quite straightforward. You can type a number, retrieve it from another column, or generate it via a calculation. All of the usual arithmetic operators (+, -, *, /) are available for performing calculations and parentheses may be used to dictate precedence, as in:

```
mysql> SELECT (37 * 59) / (78 - (8 * 6));
+---------------------------+
| (37 * 59) / (78 - (8 * 6)) |
+---------------------------+
```

```
|                   72.77 |
+--------------------------+
1 row in set (0.00 sec)
```

As was mentioned in Chapter 2, the main concern when storing numeric data is that numbers might be rounded (sometimes severely) if they are larger than the specified size for a numeric column. For example, the number 999.99 will be rounded to 99.9 if stored in a column defined as float(3,1).

Performing Arithmetic Functions

Most of the built-in numeric functions are used for specific arithmetic purposes, such as determining the square root of a number. Table 7-1 lists some of the common numeric functions that take a single numeric argument and return a number.

Table 7-1. Single-argument numeric functions

Function name	Description
Acos(x)	Calculates the arc cosine of x
Asin(x)	Calculates the arc sine of x
Atan(x)	Calculates the arc tangent of x
Cos(x)	Calculates the cosine of x
Cot(x)	Calculates the cotangent of x
Exp(x)	Calculates e^x
Ln(x)	Calculates the natural log of x
Sin(x)	Calculates the sine of x
Sqrt(x)	Calculates the square root of x
Tan(x)	Calculates the tangent of x

These functions perform very specific tasks, and I will refrain from showing examples for these functions (if you don't recognize a function by name or description, then you probably don't need it). Other numeric functions used for calculations, however, are a bit more flexible and deserve some explanation.

For example, the modulo operator, which calculates the remainder when one number is divided into another number, is implemented in MySQL and Oracle Database via the mod() function. The following example calculates the remainder when 4 is divided into 10:

```
mysql> SELECT MOD(10,4);
+-----------+
| MOD(10,4) |
+-----------+
|         2 |
+-----------+
1 row in set (0.02 sec)
```

While the `mod( )` function is typically used with integer arguments, with MySQL 4.1.7 and above you can also use real numbers, as in:

```
mysql> SELECT MOD(22.75, 5);
+---------------+
| MOD(22.75, 5) |
+---------------+
|          2.75 |
+---------------+
1 row in set (0.02 sec)
```

 SQL Server does not have a `mod( )` function. Instead, the operator `%` is used for finding remainders. The expression `10 % 4` will therefore yield the value 2.

Another numeric function that takes two numeric arguments is the `pow( )` function (or `power( )` if you are using Oracle Database or SQL Server), which returns one number raised to the power of a second number, as in:

```
mysql> SELECT POW(2,8);
+----------+
| POW(2,8) |
+----------+
|      256 |
+----------+
1 row in set (0.03 sec)
```

Thus, `pow(2,8)` is the MySQL equivalent of specifying 2^8. Since computer memory is allocated in chunks of 2^x bytes, the `pow( )` function can be a handy way to determine the exact number of bytes in a certain amount of memory:

```
mysql> SELECT POW(2,10) kilobyte, POW(2,20) megabyte,
    ->        POW(2,30) gigabyte, POW(2,40) terabyte;
+----------+----------+------------+----------------+
| kilobyte | megabyte | gigabyte   | terabyte       |
+----------+----------+------------+----------------+
|     1024 |  1048576 | 1073741824 | 1099511627776  |
+----------+----------+------------+----------------+
1 row in set (0.00 sec)
```

I don't know about you, but I find it easier to remember that a gigabyte is 2^{30} bytes than to remember the number 1,073,741,824.

Controlling Number Precision

When working with floating-point numbers, you may not always want to interact with or display a number with its full precision. For example, you may store monetary transaction data with a precision to six decimal places, but you might want to round to the nearest hundredth for display purposes. Four functions that are useful when limiting the precision of floating-point numbers: `ceil( )`, `floor( )`, `round( )`, and truncate( ). All three servers include these functions, although Oracle Database includes trunc( ) instead of truncate( ), and SQL Server includes ceiling( ) instead of ceil( ).

The ceil( ) and floor( ) functions are used to round either up or down to the closest integer, as demonstrated by the following:

```
mysql> SELECT CEIL(72.445), FLOOR(72.445);
+--------------+---------------+
| CEIL(72.445) | FLOOR(72.445) |
+--------------+---------------+
|           73 |            72 |
+--------------+---------------+
1 row in set (0.06 sec)
```

Thus, any number in between 72 and 73 will be evaluated as 73 by the ceil( ) function and 72 by the floor( ) function. Remember that ceil( ) will round up even if the decimal portion of a number is very small, and floor( ) will round down even if the decimal portion is quite significant, as in:

```
mysql> SELECT CEIL(72.000000001), FLOOR(72.999999999);
+--------------------+---------------------+
| CEIL(72.000000001) | FLOOR(72.999999999) |
+--------------------+---------------------+
|                 73 |                  72 |
+--------------------+---------------------+
1 row in set (0.00 sec)
```

If this is a bit too severe for your application, you can use the round( ) function to round up or down from the *midpoint* between two integers, as in:

```
mysql> SELECT ROUND(72.49999), ROUND(72.5), ROUND(72.50001);
+-----------------+-------------+-----------------+
| ROUND(72.49999) | ROUND(72.5) | ROUND(72.50001) |
+-----------------+-------------+-----------------+
|              72 |          72 |              73 |
+-----------------+-------------+-----------------+
1 row in set (0.00 sec)
```

Using round( ), any number whose decimal portion is more than halfway between two integers will be rounded up, whereas the number will be rounded down if the decimal portion is anything up to halfway between the two integers.

Most of the time, you will want to keep at least some part of the decimal portion of a number rather than rounding to the nearest integer; the round( ) function allows an optional second argument to specify how many digits to the right of the decimal place to round to. The next example shows how the second argument can be used to round the number 72.0909 to one, two, and three decimal places:

```
mysql> SELECT ROUND(72.0909, 1), ROUND(72.0909, 2), ROUND(72.0909, 3);
+-------------------+-------------------+-------------------+
| ROUND(72.0909, 1) | ROUND(72.0909, 2) | ROUND(72.0909, 3) |
+-------------------+-------------------+-------------------+
```

```
|                   72.1 |            72.09 |          72.091 |
+-------------------+----------------------+------------------+
1 row in set (0.00 sec)
```

Like the round( ) function, the truncate( ) function allows an optional second argument to specify the number of digits to the right of the decimal, but truncate( ) simply discards the unwanted digits without rounding. The next example shows how the number 72.0909 would be truncated to one, two, and three decimal places:

```
mysql> SELECT TRUNCATE(72.0909, 1), TRUNCATE(72.0909, 2),
    ->        TRUNCATE(72.0909, 3);
+----------------------+----------------------+----------------------+
| TRUNCATE(72.0909, 1) | TRUNCATE(72.0909, 2) | TRUNCATE(72.0909, 3) |
+----------------------+----------------------+----------------------+
|                 72.0 |                72.09 |               72.090 |
+----------------------+----------------------+----------------------+
1 row in set (0.00 sec)
```

 SQL Server does not include a truncate( ) function. Instead, the round( ) function allows for an optional third argument which, if present and nonzero, calls for the number to be truncated rather than rounded.

Both truncate( ) and round( ) also allow a *negative* value for the second argument, meaning that numbers to the *left* of the decimal place are truncated or rounded. This might seem like a strange thing to do at first, but there are valid applications. For example, you might sell a product that can only be purchased in units of ten. If a customer were to order 17 units, you could choose from one of the following methods to modify the customer's order quantity:

```
mysql> SELECT ROUND(17, -1), TRUNCATE(17, -1);
+---------------+------------------+
| ROUND(17, -1) | TRUNCATE(17, -1) |
+---------------+------------------+
|            20 |               10 |
+---------------+------------------+
1 row in set (0.00 sec)
```

If the product in question is thumbtacks, then it might not make much difference to your bottom line whether you sold the customer 10 or 20 thumbtacks when only 17 were requested; if you are selling Rolex watches, however, your business may fare better by rounding.

Handling Signed Data

If you are working with numeric columns that allow negative values (in Chapter 2, I showed how a numeric column may be labeled *unsigned*, meaning that only positive numbers are allowed), there are several numeric functions that might be of use. Let's say, for example, that you are asked to generate a report showing the current status of each bank account. The following query returns three columns useful for generating the report:

```
mysql> SELECT account_id, SIGN(avail_balance), ABS(avail_balance)
    -> FROM account;
+------------+---------------------+--------------------+
| account_id | SIGN(avail_balance) | ABS(avail_balance) |
+------------+---------------------+--------------------+
|          1 |                   1 |            1057.75 |
|          2 |                   1 |             500.00 |
|          3 |                   1 |            3000.00 |
|          4 |                   1 |            2258.02 |
|          5 |                   1 |             200.00 |
| ...                                                   |
|         19 |                   1 |            1500.00 |
|         20 |                   1 |           23575.12 |
|         21 |                   0 |               0.00 |
|         22 |                   1 |            9345.55 |
|         23 |                   1 |           38552.05 |
|         24 |                   1 |           50000.00 |
+------------+---------------------+--------------------+
24 rows in set (0.00 sec)
```

The second column uses the sign( ) function to return a –1 if the account balance is negative, 0 if the account balance is zero, and 1 if the account balance is positive. The third column returns the absolute value of the account balance via the abs( ) function.

Working with Temporal Data

Of the three types of data discussed in this chapter (character, numeric, and temporal), temporal data is the most involved when it comes to data generation and manipulation. Some of the complexity of temporal data is caused by the myriad ways in which a single date and time can be described. For example, the date on which I wrote this paragraph can be described in all of the following ways:

- Saturday, March 19, 2005
- 3/19/2005 2:14:56 P.M. EST
- 3/19/2005 19:14:56 GMT
- 0782005 (Julian format)
- Star date [–4] 82213.47 14:14:56 (*Star Trek* format)

While some of these differences are purely a matter of formatting, most of the complexity has to do with your frame of reference, which will be explored in the next section.

Dealing with Time Zones

Because people around the world prefer that noon coincides roughly with the sun's peak at their location, there has never been a serious attempt to coerce everyone to use a universal clock. Instead, the world has been sliced into 24 imaginary sections, called *time zones*; within a particular time zone, everyone agrees on the current time, whereas people in different time zones do not. While this seems simple enough, some geographic regions shift their time by one hour twice a year (implementing what is known as *Daylight Savings Time*), and some do not, so the time difference between two points on Earth might be four hours for half the year and five hours for the other half of the year. Even within a single time zone, different regions may or may not adhere to Daylight Savings Time, causing different clocks in the same time zone to agree for half the year but be one hour different for the rest of the year.

While the computer age has exacerbated the issue, people have been dealing with time zone differences since the early days of naval exploration. To ensure a common point of reference for timekeeping, fifteenth-century navigators set their clocks to the time of day in Greenwich, England. This became known as *Greenwich Mean Time*, or GMT. All other time zones can be described by the number of hours difference from GMT; for example, the time zone for the Eastern United States, known as *Eastern Standard Time,* can be described as GMT −5:00, or five hours earlier than GMT.

Today, we use a variation of GMT called *coordinated universal time*, or UTC, which is based on an atomic clock (or, to be more precise, the average time of 200 atomic clocks in 50 locations worldwide, which is referred to as *universal time*). Both SQL Server and MySQL provide functions that will return the current UTC timestamp (getutcdate( ) for SQL Server and utc_timestamp( ) for MySQL).

Most database servers default to the time zone setting of the server on which it resides and provide tools for modifying the time zone if needed. For example, a database used to store stock exchange transactions from around the world would generally be configured to use UTC time, whereas a database used to store transactions at a particular retail establishment might use the server's time zone.

MySQL keeps two different time zone settings: a global time zone, and a session time zone, which may be different for each user logged in to a database. You can see both settings via the following query:

```
mysql> SELECT @@global.time_zone, @@session.time_zone;
+--------------------+---------------------+
| @@global.time_zone | @@session.time_zone |
+--------------------+---------------------+
| SYSTEM             | SYSTEM              |
+--------------------+---------------------+
1 row in set (0.00 sec)
```

A value of system tells you that the server is using the time zone setting from the server on which the database resides.

If you are sitting at a computer in Zurich, Switzerland and open a session across the network to a MySQL server situated in New York, you may want to change the time zone setting for your session, which you can do via the following command:

```
mysql> SET time_zone = 'Europe/Zurich';
Query OK, 0 rows affected (0.18 sec)
```

If you check the time zone settings again, you will see the following:

```
mysql> SELECT @@global.time_zone, @@session.time_zone;
+--------------------+---------------------+
| @@global.time_zone | @@session.time_zone |
+--------------------+---------------------+
| SYSTEM             | Europe/Zurich       |
+--------------------+---------------------+
1 row in set (0.00 sec)
```

All dates displayed in your session will now conform to Zurich time.

 Oracle Database users can change the time zone setting for a session via the following command:

```
ALTER SESSION TIMEZONE = 'Europe/Zurich'
```

Temporal Data Generation

Temporal data may be generated via any of the following means:

- Copying data from an existing date, datetime, or time column
- Executing a built-in function that returns a date, datetime, or time
- Building a string representation of the temporal data to be evaluated by the server

To use the latter method, you will need to have an understanding of the various components used in formatting dates.

String representations of temporal data

Table 2-5 presented the more popular date components; to refresh your memory, Table 7-2 shows these same components.

Loading MySQL Time Zone Data

If you are running the MySQL server on a Windows platform, you will need to load time zone data manually before you can set global or session time zones. To do so, you need to follow these steps:

1. Download the time zone data from *http://dev.mysql.com/downloads/timezones.html*.
2. Shut down your MySQL server.
3. Extract the files from the downloaded zip file (in my case, the file was called *timezone-2004e.zip*) and place them in your MySQL installation directory under */data/mysql* (the full path for my installation was */Program Files/MySQL/MySQL Server 4.1/data/mysql*).
4. Restart your MySQL server

To look at the time zone data, change to the mysql database via the use mysql command, and execute the following query:

```
mysql> SELECT name FROM time_zone_name;
+--------------------------------+
| name                           |
+--------------------------------+
| Africa/Abidjan                 |
| Africa/Accra                   |
| Africa/Addis_Ababa             |
| Africa/Algiers                 |
| Africa/Asmera                  |
| Africa/Bamako                  |
| Africa/Bangui                  |
| Africa/Banjul                  |
| Africa/Bissau                  |
| Africa/Blantyre                |
| Africa/Brazzaville             |
| Africa/Bujumbura               |
...
| US/Alaska                      |
| US/Aleutian                    |
| US/Arizona                     |
| US/Central                     |
| US/East-Indiana                |
| US/Eastern                     |
| US/Hawaii                      |
| US/Indiana-Starke              |
| US/Michigan                    |
| US/Mountain                    |
| US/Pacific                     |
| US/Samoa                       |
| UTC                            |
```

—continued—

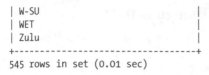

```
| W-SU                              |
| WET                               |
| Zulu                              |
+-----------------------------------+
545 rows in set (0.01 sec)
```

To change your time zone setting, choose one of the names from the previous query that best matches your location.

Table 7-2. Date format components

Component	Definition	Range
YYYY	Year, including century	1000 to 9999
MM	Month	01 (January) to 12 (December)
DD	Day	01 to 31
HH	Hour	00 to 23
HHH	Hours (elapsed)	–838 to 838
MI	Minute	00 to 59
SS	Second	00 to 59

To build a string that can be interpreted by the server as a date, datetime, or time, you need to put the various components together in the order shown in Table 7-3.

Table 7-3. Required date components

Type	Default format
Date	YYYY-MM-DD
Datetime	YYYY-MM-DD HH:MI:SS
Timestamp	YYYY-MM-DD HH:MI:SS
Time	HHH:MI:SS

Thus, to populate a datetime column with 3:30 P.M. on March 27, 2005, you will need to build the following string:

```
'2005-03-27 15:30:00'
```

If the server is expecting a datetime value, such as when updating a datetime column or when calling a built-in function that takes a datetime argument, you can provide a properly formatted string with the required date components, and the server will do the conversion for you. For example, here's a statement used to modify the date of a bank transaction:

```
UPDATE transaction
SET txn_date = '2005-03-27 15:30:00'
WHERE txn_id = 99999;
```

The server determines that the string provided in the set clause must be a datetime value, since the string is being used to populate a datetime column. Therefore, the server will attempt to convert the string for you by parsing the string into the six components (year, month, day, hour, minute, second) included in the default datetime format.

String-to-date conversions

If the server is *not* expecting a datetime value, you need to tell the server to convert the string to a datetime. For example, here is a simple query that returns a datetime value using the cast( ) function:

```
mysql> SELECT CAST('2005-03-27 15:30:00' AS DATETIME);
+------------------------------------------+
| CAST('2005-03-27 15:30:00' AS DATETIME)  |
+------------------------------------------+
| 2005-03-27 15:30:00                      |
+------------------------------------------+
1 row in set (0.00 sec)
```

The cast( ) function will be covered at the end of this chapter. While this example demonstrates how to build datetime values, the same logic applies to the date and time types as well. The following query uses the cast( ) function to generate a date value and a time value:

```
mysql> SELECT CAST('2005-03-27' AS DATE) date_field,
    ->   CAST('108:17:57' AS TIME) time_field;
+------------+------------+
| date_field | time_field |
+------------+------------+
| 2005-03-27 | 108:17:57  |
+------------+------------+
1 row in set (0.00 sec)
```

You may, of course, explicitly convert your strings even when the server is expecting a date, datetime, or time value, rather than letting the server do an implicit conversion.

When strings are converted to temporal values, whether explicitly or implicitly, you must provide all of the date components in the required order. While some servers are quite strict regarding the date format, the MySQL server is quite lenient about the separators used between the components. For example, MySQL will accept all of the following strings as valid representations of 3:30 P.M. on March 27, 2005:

```
'2005-03-27 15:30:00'
'2005/03/27 15:30:00'
'2005,03,27,15,30,00'
'20050327153000'
```

While this gives you a bit more flexibility, you may find yourself trying to generate a temporal value *without* the default date components; the next section will demonstrate a built-in function that is far more flexible than the cast( ) function.

Functions for generating dates

If you need to generate temporal data from a string, and the string is not in the proper form to use the cast( ) function, you can use a built-in function that allows you to provide a format string along with the date string. MySQL includes the str_to_date( ) function for this purpose. Say, for example, that you pull the string "March 27, 2005" from a file and need to use it to update a date column. Since the string is not in the required YYYY-MM-DD format, you can use str_to_date( ) instead of reformatting the string so that you can use the cast( ) function, as in:

```
UPDATE individual
SET birth_date = STR_TO_DATE('March 27, 2005', '%M %d, %Y')
WHERE cust_id = 9999;
```

The second argument in the call to str_to_date( ) defines the format of the date string, with, in this case, a month name (%M), a numeric day (%d), and a four-digit numeric year (%Y). While there are over 30 recognized format components, Table 7-4 defines the dozen or so most commonly used components.

Table 7-4. Date format components

Format component	Description
%M	Month name (January to December)
%m	Month numeric (01 to 12)
%d	Day numeric (01 to 31)
%j	Day of year (001 to 366)
%W	Weekday name (Sunday to Saturday)
%Y	Year, four-digit numeric
%y	Year, two-digit numeric
%H	Hour (00 to 23)
%h	Hour (01 to 12)
%i	Minutes (00 to 59)
%s	Seconds (00 to 59)
%f	Microseconds (000000 to 999999)
%p	A.M. or P.M.

The str_to_date( ) function returns a datetime, date, or time value depending on the contents of the format string. For example, if the format string includes only %H, %i, and %s, then a time value will be returned.

> Oracle Database users can use the to_date( ) function in the same manner as MySQL's str_to_date( ) function.

If you are trying to generate the *current* date/time, then you won't need to build a string, because the following built-in functions will access the system clock and return the current date and/or time as a string for you:

```
mysql> SELECT CURRENT_DATE( ), CURRENT_TIME( ), CURRENT_TIMESTAMP( );
+----------------+----------------+---------------------+
| CURRENT_DATE( ) | CURRENT_TIME( ) | CURRENT_TIMESTAMP( ) |
+----------------+----------------+---------------------+
| 2005-03-20      | 22:15:56        | 2005-03-20 22:15:56 |
+----------------+----------------+---------------------+
1 row in set (0.00 sec)
```

The values returned by these functions are in the default format for the temporal type being returned. Oracle Database includes current_date( ) and current_timestamp( ) but not current_time( ), and SQL Server includes only the current_timestamp( ) function.

Temporal Data Manipulation

This section explores the built-in functions that take date arguments and return dates, strings, or numbers.

Temporal functions that return dates

Many of the built-in temporal functions take one date as an argument and return another date. MySQL's date_add( ) function, for example, allows you to add any kind of interval (i.e., days, months, years) to a specified date to generate another date. Here's an example that demonstrates how to add five days to the current date:

```
mysql> SELECT DATE_ADD(CURRENT_DATE( ), INTERVAL 5 DAY);
+------------------------------------------+
| DATE_ADD(CURRENT_DATE( ), INTERVAL 5 DAY) |
+------------------------------------------+
| 2005-03-26                               |
+------------------------------------------+
1 row in set (0.00 sec)
```

The second argument comprises three elements: the interval keyword, the desired quantity, and the type of interval. Table 7-5 shows some of the commonly used interval types.

Table 7-5. Common interval types

Interval name	Description
Second	Number of seconds
Minute	Number of minutes
Hour	Number of hours
Day	Number of days
Month	Number of months

Table 7-5. Common interval types (continued)

Interval name	Description
Year	Number of years
Minute_second	Number of minutes and seconds, separated by ":"
Hour_second	Number of hours, minutes, and seconds, separated by ":"
Year_month	Number of years and months, separated by "-"

While the first six types listed in Table 7-5 are pretty straightforward, the last three types require a bit more explanation since they have multiple elements. For example, if you are told that transaction ID 9999 actually occurred 3 hours, 27 minutes, and 11 seconds later than what was posted to the transaction table, you can fix it via the following:

```
UPDATE transaction
SET txn_date = DATE_ADD(txn_date, INTERVAL '3:27:11' HOUR_SECOND)
WHERE txn_id = 9999;
```

In this example, the date_add( ) function takes the value in the txn_date column, adds 3 hours, 27 minutes, and 11 seconds to it, and uses the resulting value to modify the txn_date column.

Or, if you work in HR and found out that employee ID 4789 claimed to be younger than he actually is, you could add 9 years and 11 months to his birth date, as in:

```
UPDATE employee
SET birth_date = DATE_ADD(birth_date, INTERVAL '9-11' YEAR_MONTH)
WHERE emp_id = 4789;
```

For SQL Server users, the previous example could be accomplished using the dateadd( ) function:

```
UPDATE employee
SET birth_date =
  DATEADD(MONTH, 119, birth_date)
WHERE emp_id = 4789
```

SQL Server doesn't have combined intervals (i.e., year_month), so I converted 9 years, 11 months to 119 months.

Oracle Database users can use the add_months( ) function for this example, as in:

```
UPDATE employee
SET birth_date = ADD_MONTHS(birth_date, 119)
WHERE emp_id = 4789;
```

There are some cases where you want to add an interval to a date, and you know where you want to arrive but not how many days it takes to get there. For example, let's say that a bank customer logs onto the online banking system and schedules a transfer for the end of the month. Rather than writing some code that figures out what month you are currently in and looks up the number of days in that month, you can call the last_day( ) function, which does the work for you (both MySQL and Oracle Database include the last_day( ) function; SQL Server has no comparable function). If the customer asks for the transfer on March 25, 2005, you could find the last day of March via the following:

```
mysql> SELECT LAST_DAY('2005-03-25');
+------------------------+
| LAST_DAY('2005-03-25') |
+------------------------+
| 2005-03-31             |
+------------------------+
1 row in set (0.04 sec)
```

Whether you provide a date or datetime value, the last_day( ) function always returns a date. While this function may not seem like an enormous time-saver, the underlying logic can be tricky if you're trying to find the last day of February and need to figure out whether the current year is a leap year or not.

Another temporal function that returns a date is one that converts a datetime value from one time zone to another. For this purpose, MySQL includes the convert_tz( ) function and Oracle Database includes the new_time( ) function. If I want to convert my current local time to UTC, for example, I could do the following:

```
mysql> SELECT CURRENT_TIMESTAMP() current_est,
    ->    CONVERT_TZ(CURRENT_TIMESTAMP(), 'US/Eastern', 'UTC') current_utc;
+---------------------+---------------------+
| current_est         | current_utc         |
+---------------------+---------------------+
| 2005-04-18 21:23:25 | 2005-04-19 01:23:25 |
+---------------------+---------------------+
1 row in set (0.50 sec)
```

This function comes in handy when receiving dates in a different time zone than what is stored in your database.

Temporal functions that return strings

Most of the temporal functions that return string values are used to extract a portion of a date or time. For example, MySQL includes the dayname( ) function to determine which day of the week a certain date falls on, as in:

```
mysql> SELECT DAYNAME('2005-03-22');
+-----------------------+
| DAYNAME('2005-03-22') |
+-----------------------+
| Tuesday               |
+-----------------------+
1 row in set (0.12 sec)
```

There are many such functions included with MySQL for extracting information from date values, but I recommend that you use the extract( ) function instead, since it's easier to remember a few variations of one function than to remember a

dozen different functions. Additionally, the extract( ) function is part of the SQL:
2003 standard and has been implemented by Oracle Database as well as MySQL.

The extract( ) function uses the same interval types as the date_add( ) function (see
Table 7-5) to define which element of the date that interests you. For example, if you
want to extract just the year portion of a datetime value, you can do the following:

```
mysql> SELECT EXTRACT(YEAR FROM '2005-03-22 22:19:05');
+-----------------------------------------+
| EXTRACT(YEAR FROM '2005-03-22 22:19:05') |
+-----------------------------------------+
|                                    2005 |
+-----------------------------------------+
1 row in set (0.02 sec)
```

 SQL Server doesn't include an implementation of extract( ), but it
does include the datepart( ) function. Here's how you would extract
the year from a datetime value using datepart( ):

SELECT DATEPART(YEAR, GETDATE( ))

Temporal functions that return numbers

Earlier in this chapter, I showed you a function used to add a given interval to a date
value, thus generating another date value. Another common activity when working
with dates is to take two date values and determine the number of intervals (days,
weeks, years) *between* the two dates. For this purpose, MySQL includes the function
datediff( ), which returns the number of full days between two dates. For example,
if I want to know the number of days that my kids will be out of school this sum-
mer, I can do the following:

```
mysql> SELECT DATEDIFF('2005-09-05', '2005-06-22');
+-------------------------------------+
| DATEDIFF('2005-09-05', '2005-06-22') |
+-------------------------------------+
|                                  75 |
+-------------------------------------+
1 row in set (0.00 sec)
```

Thus, I will have to endure 75 days of poison ivy, mosquito bites, and scraped knees
before the kids are safely back at school. The datediff( ) function ignores the time of
day in its arguments. Even if I include a time-of-day, setting for the first date of one
second until midnight, and the second date of one second after midnight, those times
will have no effect on the calculation:

```
mysql> SELECT DATEDIFF('2005-09-05 23:59:59', '2005-06-22 00:00:01');
+-------------------------------------------------------+
| DATEDIFF('2005-09-05 23:59:59', '2005-06-22 00:00:01') |
+-------------------------------------------------------+
|                                                    75 |
+-------------------------------------------------------+
1 row in set (0.00 sec)
```

If I switch the arguments and have the earlier date first, `datediff( )` will return a negative number, as in:

```
mysql> SELECT DATEDIFF('2005-06-22', '2005-09-05');
+--------------------------------------+
| DATEDIFF('2005-06-22', '2005-09-05') |
+--------------------------------------+
|                                  -75 |
+--------------------------------------+
```

 SQL Server also includes the `datediff( )` function, but it is more flexible than the MySQL implementation in that you can specify the interval type (i.e., year, month, day, hour) instead of counting only the number of days between two dates. Here's how SQL Server would accomplish the previous example:

```
SELECT DATEDIFF(DAY, '2005-06-22',
   '2005-09-05')
```

Oracle Database allows you to determine the number of days between two dates simply by subtracting one date from another.

Conversion Functions

Earlier in this chapter, I showed you how to use the `cast( )` function to convert a string to a `datetime` value. While every database server includes a number of proprietary functions used to convert data from one type to another, I recommend using the `cast( )` function, which is included in the SQL:2003 standard and has been implemented by MySQL, Oracle Database, and Microsoft SQL Server.

To use `cast( )`, you provide a value or expression, the as keyword, and the type you wish the value converted to. Here's an example that converts a string to an integer:

```
mysql> SELECT CAST('1456328' AS SIGNED INTEGER);
+-----------------------------------+
| CAST('1456328' AS SIGNED INTEGER) |
+-----------------------------------+
|                           1456328 |
+-----------------------------------+
1 row in set (0.01 sec)
```

When converting a string to a number, the `cast( )` function will attempt to convert the entire string from left to right; if any nonnumeric characters are found in the string, then the conversion halts without an error. Consider the following example:

```
mysql> SELECT CAST('999ABC111' AS UNSIGNED INTEGER);
+---------------------------------------+
| CAST('999ABC111' AS UNSIGNED INTEGER) |
+---------------------------------------+
|                                   999 |
+---------------------------------------+
1 row in set (0.00 sec)
```

In this case, the first three digits of the string are converted, whereas the rest of the string is discarded, resulting in a value of 999.

If you are converting a string to a date, time, or datetime value, then you will need to stick with the default formats for each type, since you can't provide the cast( ) function with a format string. If your date string is not in the default format (i.e., YYYY-MM-DD HH:MI:SS for datetime types), then you will need to resort to using another function, such as MySQL's str_to_date( ) function described earlier in the chapter.

Exercises

These exercises are designed to test your understanding of some of the built-in functions shown in this chapter. See Appendix C for the answers.

7-1

Write a query that returns the 17th through 25th characters of the string "Please find the substring in this string."

7-2

Write a query that returns the absolute value and sign (–1, 0, or 1) of the number –25.76823. Also return the number rounded to the nearest hundredth.

7-3

Write a query to return just the month portion of the current date.

CHAPTER 8

Grouping and Aggregates

Data is generally stored at the lowest level of granularity needed by any of a database's users; if Chuck in accounting needs to look at individual customer transactions, then there needs to be a table in the database that stores individual transactions. That doesn't mean, however, that all users must deal with the data as it is stored in the database. The focus of this chapter will be on how data can be grouped and aggregated to allow users to interact with data at some higher level of granularity than what is stored in the database.

Grouping Concepts

Sometimes, you will want to find trends in your data that will require the database server to cook the data a bit before you can generate the results you are looking for. For example, let's say that you are in charge of operations at the bank, and you would like to find out how many accounts are being opened by each bank teller. You could issue a simple query to look at the raw data:

```
mysql> SELECT open_emp_id
    -> FROM account;
+-------------+
| open_emp_id |
+-------------+
|           1 |
|           1 |
|           1 |
|           1 |
|           1 |
|           1 |
|           1 |
|           1 |
|          10 |
|          10 |
|          10 |
|          10 |
|          10 |
```

```
|          10 |
|          10 |
|          13 |
|          13 |
|          13 |
|          16 |
|          16 |
|          16 |
|          16 |
|          16 |
|          16 |
+-------------+
24 rows in set (0.01 sec)
```

With only 24 rows in the account table, it is relatively easy to see that four different employees opened accounts and that employee ID 16 has opened six accounts; however, if the bank has dozens of employees and thousands of accounts, this approach would prove tedious and error prone.

Instead, you can ask the database server to group the data for you by using the group by clause. Here's the same query but employing a group by clause to group the account data by employee ID:

```
mysql> SELECT open_emp_id
    -> FROM account
    -> GROUP BY open_emp_id;
+-------------+
| open_emp_id |
+-------------+
|           1 |
|          10 |
|          13 |
|          16 |
+-------------+
4 rows in set (0.00 sec)
```

The result set contains one row for each distinct value in the open_emp_id column, resulting in four rows instead of the full 24 rows. The reason for the smaller result set is that each of the four employees opened more than one account. To see how many accounts each teller opened, you can use an *aggregate function* in the select clause to count the number of rows in each group:

```
mysql> SELECT open_emp_id, COUNT(*) how_many
    -> FROM account
    -> GROUP BY open_emp_id;
+-------------+----------+
| open_emp_id | how_many |
+-------------+----------+
|           1 |        8 |
|          10 |        7 |
|          13 |        3 |
|          16 |        6 |
+-------------+----------+
4 rows in set (0.00 sec)
```

The aggregate function count( ) counts the number of rows in each group, and the asterisk tells the server to count everything in the group. Using the combination of a group by clause and the count( ) aggregate function, you are able to generate exactly the data needed to answer the business question without having to look at the raw data.

When grouping data, you may need to filter out undesired data from your result set based on groups of data rather than based on the raw data. Since the group by clause runs *after* the where clause has been evaluated, you cannot add filter conditions to your where clause for this purpose. For example, here's an attempt to filter out any cases where an employee has opened fewer than five accounts:

```
mysql> SELECT open_emp_id, COUNT(*) how_many
    -> FROM account
    -> WHERE COUNT(*) > 4
    -> GROUP BY open_emp_id, product_cd;
ERROR 1111 (HY000): Invalid use of group function
```

You cannot refer to the aggregate function count(*) in your where clause, because the groups have not yet been generated at the time the where clause is evaluated. Instead, you can put your group filter conditions in the having clause. Here's what the query would look like using having:

```
mysql> SELECT open_emp_id, COUNT(*) how_many
    -> FROM account
    -> GROUP BY open_emp_id
    -> HAVING COUNT(*) > 4;
+-------------+----------+
| open_emp_id | how_many |
+-------------+----------+
|           1 |        8 |
|          10 |        7 |
|          16 |        6 |
+-------------+----------+
3 rows in set (0.00 sec)
```

Because those groups containing fewer than five members have been filtered out via the having clause, the result set now contains only those employees who have opened five or more accounts.

Aggregate Functions

Aggregate functions perform a specific operation over all rows in a group. Although every database server has its own set of specialty aggregate functions, the common aggregate functions implemented by all major servers include:

Max( )
> Returns the maximum value within a set

Min( )
> Return the minimum value within a set

Avg( )
   Returns the average value within a set

Sum( )
   Returns the sum of the values in a set

Count( )
   Returns the number of values in a set

Here's a query that uses all of the common aggregate functions to analyze the available balances for all checking accounts:

```
mysql> SELECT MAX(avail_balance) max_balance,
    ->   MIN(avail_balance) min_balance,
    ->   AVG(avail_balance) avg_balance,
    ->   SUM(avail_balance) tot_balance,
    ->   COUNT(*) num_accounts
    -> FROM account
    -> WHERE product_cd = 'CHK';
+-------------+-------------+-------------+-------------+--------------+
| max_balance | min_balance | avg_balance | tot_balance | num_accounts |
+-------------+-------------+-------------+-------------+--------------+
|    38552.05 |      122.37 | 7300.800985 |    73008.01 |           10 |
+-------------+-------------+-------------+-------------+--------------+
1 row in set (0.09 sec)
```

The results from this query tell you that, across the ten checking accounts in the account table, there is a maximum balance of $38,552.05, a minimum balance of $122.37, an average balance of $7,300.80, and a total balance across all ten accounts of $73,008.01. Hopefully, this gives you an appreciation for the role of these aggregate functions; the next subsections will further clarify how these functions can be utilized.

Implicit Versus Explicit Groups

In the previous example, every value returned by the query is generated by an aggregate function, and the aggregate functions are applied across the group of rows specified by the filter condition product_cd = 'CHK'. Since there is no group by clause, there is a single, *implicit* group (all rows returned by the query).

In most cases, however, you will want to retrieve additional columns along with columns generated by aggregate functions. What if, for example, you wanted to extend the previous query to execute the same five aggregate functions for *each* product type, instead of just for checking accounts? For this query, you would want to retrieve the product_cd column along with the five aggregate functions, as in:

```
SELECT product_cd,
   MAX(avail_balance) max_balance,
   MIN(avail_balance) min_balance,
   AVG(avail_balance) avg_balance,
   SUM(avail_balance) tot_balance,
```

```
        COUNT(*) num_accounts
FROM account;
```

However, if you try to execute the query, you will receive the following error:

```
ERROR 1140 (42000): Mixing of GROUP columns (MIN(),MAX(),COUNT(),...) with no GROUP
columns is illegal if there is no GROUP BY clause
```

While it may be obvious to you that you want the aggregate functions applied to each set of products found in the account table, this query fails because you have not *explicitly* specified how the data should be grouped. Therefore, you will need to add a group by clause to specify over which group of rows the aggregates functions should be applied:

```
mysql> SELECT product_cd,
    ->    MAX(avail_balance) max_balance,
    ->    MIN(avail_balance) min_balance,
    ->    AVG(avail_balance) avg_balance,
    ->    SUM(avail_balance) tot_balance,
    ->    COUNT(*) num_accts
    -> FROM account
    -> GROUP BY product_cd;
+------------+-------------+-------------+-------------+-------------+-----------+
| product_cd | max_balance | min_balance | avg_balance | tot_balance | num_accts |
+------------+-------------+-------------+-------------+-------------+-----------+
| BUS        |     9345.55 |        0.00 | 4672.774902 |     9345.55 |         2 |
| CD         |    10000.00 |     1500.00 | 4875.000000 |    19500.00 |         4 |
| CHK        |    38552.05 |      122.37 | 7300.800985 |    73008.01 |        10 |
| MM         |     9345.55 |     2212.50 | 5681.713216 |    17045.14 |         3 |
| SAV        |      767.77 |      200.00 |  463.940002 |     1855.76 |         4 |
| SBL        |    50000.00 |    50000.00 | 50000.000000 |    50000.00 |         1 |
+------------+-------------+-------------+-------------+-------------+-----------+
6 rows in set (0.00 sec)
```

With the inclusion of the group by clause, the server knows to group together rows having the same value in the product_cd column first, and then apply the five aggregate functions to each of the six groups.

Counting Distinct Values

When using the count( ) function to determine the number of members in each group, you have your choice of counting *all* members in the group, or counting only the *distinct* values for a column across all members of the group. For example, consider the following data, which shows the employee responsible for opening each account:

```
mysql> SELECT account_id, open_emp_id
    -> FROM account
    -> ORDER BY open_emp_id;
+------------+-------------+
| account_id | open_emp_id |
+------------+-------------+
|          8 |           1 |
```

```
|          9 |            1 |
|         10 |            1 |
|         12 |            1 |
|         13 |            1 |
|         17 |            1 |
|         18 |            1 |
|         19 |            1 |
|          1 |           10 |
|          2 |           10 |
|          3 |           10 |
|          4 |           10 |
|          5 |           10 |
|         14 |           10 |
|         22 |           10 |
|          6 |           13 |
|          7 |           13 |
|         24 |           13 |
|         11 |           16 |
|         15 |           16 |
|         16 |           16 |
|         20 |           16 |
|         21 |           16 |
|         23 |           16 |
+------------+-------------+
24 rows in set (0.00 sec)
```

As you can see, multiple accounts were opened by four different employees (employee IDs 1, 10, 13, and 16). Let's say that, instead of performing a manual count, you want to create a query that counts the number of employees who have opened accounts. If you apply the count( ) function to the open_emp_id column, you will see the following results:

```
mysql> SELECT COUNT(open_emp_id)
    -> FROM account;
+--------------------+
| COUNT(open_emp_id) |
+--------------------+
|                 24 |
+--------------------+
1 row in set (0.00 sec)
```

In this case, specifying the open_emp_id column as the column to be counted doesn't generate any different results than having specified count(*). If you want to count *distinct* values in the group rather than just counting the number of rows in the group, you need to specify the distinct keyword, as in:

```
mysql> SELECT COUNT(DISTINCT open_emp_id)
    -> FROM account;
+----------------------------+
| COUNT(DISTINCT open_emp_id) |
+----------------------------+
|                          4 |
+----------------------------+
1 row in set (0.00 sec)
```

By specifying `distinct`, therefore, the `count( )` function examines the values of a column for each member of the group, rather than simply counting the number of values in the group.

Using Expressions

Along with using columns as arguments to aggregate functions, you can build expressions to use as arguments. For example, you may want to find the maximum value of pending deposits across all accounts, which is calculated by subtracting the available balance from the pending balance. You can achieve this via the following query:

```
mysql> SELECT MAX(pending_balance - avail_balance) max_uncleared
    -> FROM account;
+---------------+
| max_uncleared |
+---------------+
|        660.00 |
+---------------+
1 row in set (0.00 sec)
```

While this example uses a fairly simple expression, expressions used as arguments to aggregate functions can be as complex as needed, as long as they return a number, string, or date. In Chapter 11, I will show you how you can use case expressions with aggregate functions to determine whether a particular row should or should not be included in an aggregation.

How Nulls Are Handled

When performing aggregations, or, indeed, any type of numeric calculation, you should always consider how `null` values might affect the outcome of your calculation. To illustrate, I will build a simple table to hold numeric data and populate it with the set {1, 3, 5}:

```
mysql> CREATE TABLE number_tbl
    -> (val SMALLINT);
Query OK, 0 rows affected (0.01 sec)

mysql> INSERT INTO number_tbl VALUES (1);
Query OK, 1 row affected (0.00 sec)

mysql> INSERT INTO number_tbl VALUES (3);
Query OK, 1 row affected (0.00 sec)

mysql> INSERT INTO number_tbl VALUES (5);
Query OK, 1 row affected (0.00 sec)
```

Consider the following query, which builds a table named data consisting of three numbers, and then performs five aggregate functions on the set of numbers:

```
mysql> SELECT COUNT(*) num_rows,
    ->   COUNT(val) num_vals,
    ->   SUM(val) total,
    ->   MAX(val) max_val,
    ->   AVG(val) avg_val
    -> FROM number_tbl;
+----------+----------+-------+---------+---------+
| num_rows | num_vals | total | max_val | avg_val |
+----------+----------+-------+---------+---------+
|        3 |        3 |     9 |       5 |       3 |
+----------+----------+-------+---------+---------+
1 row in set (0.00 sec)
```

The results are as you would expect: both count(*) and count(data.num) return the value 3, sum(data.num) returns the value 9, max(data.num) returns 5, and avg(data.num) returns 3. Next, I will add a null value to the number_tbl table and run the query again:

```
mysql> INSERT INTO number_tbl VALUES (NULL);
Query OK, 1 row affected (0.01 sec)

mysql> SELECT COUNT(*) num_rows,
    ->   COUNT(val) num_vals,
    ->   SUM(val) total,
    ->   MAX(val) max_val,
    ->   AVG(val) avg_val
    -> FROM number_tbl;
+----------+----------+-------+---------+---------+
| num_rows | num_vals | total | max_val | avg_val |
+----------+----------+-------+---------+---------+
|        4 |        3 |     9 |       5 |       3 |
+----------+----------+-------+---------+---------+
1 row in set (0.00 sec)
```

Even with the addition of the null value to the table, the sum( ), max( ), and avg( ) functions all return the same values, indicating that they ignore any null values encountered. The count(*) function now returns the value 4, which is valid since the number_tbl table contains four rows, while the count(val) function still returns the value 3. The difference is that count(*) counts the number of rows and is thus unaffected by any null values contained in a row, while count(val) counts the number of *values* contained in the val column and ignores any null values encountered.

Generating Groups

People are rarely interested in looking at raw data; instead, people engaging in data analysis will want to manipulate the raw data to better suit their needs. Examples of common data manipulations include:

- Generating totals for a geographic region, such as total European sales
- Finding outliers, such as the top salesperson for 2005
- Determining frequencies, such as the number of new accounts opened for each branch

To answer these types of queries, you will need to ask the database server to group rows together by one or more columns or expressions. As you have seen already in several examples, the group by clause is the mechanism for grouping data within a query. In this section, you will see how to group data by one or more columns, how to group data using expressions, and how to generate rollups within groups.

Single-Column Grouping

Single-column groups are the simplest and most-often-used type of grouping. If you want to find the total balances for each product, for example, you need only group on the account.product_cd column, as in:

```
mysql> SELECT product_cd, SUM(avail_balance) prod_balance
    -> FROM account
    -> GROUP BY product_cd;
+------------+--------------+
| product_cd | prod_balance |
+------------+--------------+
| BUS        |      9345.55 |
| CD         |     19500.00 |
| CHK        |     73008.01 |
| MM         |     17045.14 |
| SAV        |      1855.76 |
| SBL        |     50000.00 |
+------------+--------------+
6 rows in set (0.00 sec)
```

This query generates six groups, one for each product, and then sums the available balances across all rows in each group.

Multi-Column Grouping

In some cases, you may want to generate groups that span *more* than one column. Expanding on the previous example, imagine that you want to find the total balances not just for each product, but for products and branches (i.e., what's the total balance for all checking accounts opened at the Woburn branch?). The following example shows how this can be accomplished:

```
mysql> SELECT product_cd, open_branch_id,
    ->     SUM(avail_balance) tot_balance
    -> FROM account
    -> GROUP BY product_cd, open_branch_id;
+------------+----------------+-------------+
| product_cd | open_branch_id | tot_balance |
+------------+----------------+-------------+
| BUS        |              2 |     9345.55 |
| BUS        |              4 |        0.00 |
| CD         |              1 |    11500.00 |
| CD         |              2 |     8000.00 |
| CHK        |              1 |      782.16 |
| CHK        |              2 |     3315.77 |
| CHK        |              3 |     1057.75 |
| CHK        |              4 |    67852.33 |
| MM         |              1 |    14832.64 |
| MM         |              3 |     2212.50 |
| SAV        |              1 |      767.77 |
| SAV        |              2 |      700.00 |
| SAV        |              4 |      387.99 |
| SBL        |              3 |    50000.00 |
+------------+----------------+-------------+
14 rows in set (0.00 sec)
```

This version of the query generates 14 groups, one for each combination of product and branch found in the account table. Along with adding the open_branch_id column to the select clause, I also added it to the group by clause, since open_branch_id is retrieved from a table and not generated via an aggregate function.

Grouping via Expressions

Along with using columns to group data, you can build groups based on the values generated by expressions. Consider the following query, which groups employees by the year they began working for the bank:

```
mysql> SELECT EXTRACT(YEAR FROM start_date) year,
    ->     COUNT(*) how_many
    -> FROM employee
    -> GROUP BY EXTRACT(YEAR FROM start_date);
+------+----------+
| year | how_many |
+------+----------+
| 2000 |        3 |
| 2001 |        2 |
| 2002 |        8 |
| 2003 |        3 |
| 2004 |        2 |
+------+----------+
5 rows in set (0.00 sec)
```

This query uses a fairly simple expression, which uses the extract( ) function to return only the year portion of a date, to group the rows in the employee table.

Generating Rollups

In the "Multi-Column Grouping" section earlier in the chapter, I showed an example that generates total account balances for each product and branch. Let's say, however, that along with the total balances for each product/branch combination, you also want total balances for each distinct product. You could run an additional query and merge the results, you could load the results of the query into a spreadsheet, or you could build a Perl script, Java program, or some other mechanism to take that data and perform the additional calculations. Better yet, you could use the with rollup option to have the database server do the work for you. Here's the revised query using with rollup in the group by clause:

```
mysql> SELECT product_cd, open_branch_id,
    ->   SUM(avail_balance) tot_balance
    -> FROM account
    -> GROUP BY product_cd, open_branch_id WITH ROLLUP;
+------------+----------------+-------------+
| product_cd | open_branch_id | tot_balance |
+------------+----------------+-------------+
| BUS        |              2 |     9345.55 |
| BUS        |              4 |        0.00 |
| BUS        |           NULL |     9345.55 |
| CD         |              1 |    11500.00 |
| CD         |              2 |     8000.00 |
| CD         |           NULL |    19500.00 |
| CHK        |              1 |      782.16 |
| CHK        |              2 |     3315.77 |
| CHK        |              3 |     1057.75 |
| CHK        |              4 |    67852.33 |
| CHK        |           NULL |    73008.01 |
| MM         |              1 |    14832.64 |
| MM         |              3 |     2212.50 |
| MM         |           NULL |    17045.14 |
| SAV        |              1 |      767.77 |
| SAV        |              2 |      700.00 |
| SAV        |              4 |      387.99 |
| SAV        |           NULL |     1855.76 |
| SBL        |              3 |    50000.00 |
| SBL        |           NULL |    50000.00 |
| NULL       |           NULL |   170754.46 |
+------------+----------------+-------------+
21 rows in set (0.02 sec)
```

There are now seven additional results, one for each of the six distinct products, and one for the grand total (all products combined). For the six product rollups, a null value is provided for the open_branch_id column, since the rollup is being performed across all branches. Looking at line #3 of the output, for example, you will see that there was a total of $9,345.55 deposited in BUS accounts across all branches. For the grand total row, a null value is provided for both the product_cd and open_branch_id columns; the last line of output shows a total of $170,754.46 across all products and branches.

 If you are using Oracle Database, you need to use a slightly different syntax to indicate that you want a rollup performed. The group by clause for the previous query would look as follows when using Oracle:

```
GROUP BY ROLLUP(product_cd, open_branch_id)
```

The advantage of this syntax is that it allows you to perform rollups on a subset of the columns in the group by clause. If you are grouping by columns a, b, and c, for example, you could indicate that the server should perform rollups on only b and c via the following:

```
GROUP BY a, ROLLUP(b, c)
```

If, along with totals by product, you also want to calculate totals per branch, then you can use the with cube option, which generates summary rows for *all* possible combinations of the grouping columns. Unfortunately, with cube is not available in Version 4.1 of MySQL, but it is available with SQL Server and with Oracle Database. Here's an example using with cube, but I have removed the mysql> prompt to show that the query cannot yet be performed with MySQL:

```
SELECT product_cd, open_branch_id,
  SUM(avail_balance) tot_balance
FROM account
GROUP BY product_cd, open_branch_id WITH CUBE;
```

product_cd	open_branch_id	tot_balance
NULL	NULL	170754.46
NULL	1	27882.57
NULL	2	21361.32
NULL	3	53270.25
NULL	4	68240.32
BUS	2	9345.55
BUS	4	0.00
BUS	NULL	9345.55
CD	1	11500.00
CD	2	8000.00
CD	NULL	19500.00
CHK	1	782.16
CHK	2	3315.77
CHK	3	1057.75
CHK	4	67852.33
CHK	NULL	73008.01
MM	1	14832.64
MM	3	2212.50
MM	NULL	17045.14
SAV	1	767.77

```
| SAV         |            2 |      700.00 |
| SAV         |            4 |      387.99 |
| SAV         |         NULL |     1855.76 |
| SBL         |            3 |    50000.00 |
| SBL         |         NULL |    50000.00 |
+-------------+--------------+-------------+
25 rows in set (0.02 sec)
```

Using with cube generates four more rows than the with rollup version of the query, one for each of the four branch IDs. Similar to with rollup, null values are placed in the product_cd column to indicate that a branch summary is being performed.

 Once again, if you are using Oracle Database, you need to use a slightly different syntax to indicate that you want a cube operation performed. The group by clause for the previous query would look as follows when using Oracle:

```
GROUP BY CUBE(product_cd, open_branch_id)
```

Group Filter Conditions

In Chapter 4, you were introduced to various types of filter conditions and shown how they might be used in the where clause. When grouping data, you also have the ability to apply filter conditions to the data *after* the groups have been generated. The having clause is where these types of filter conditions should be placed. Consider the following example:

```
mysql> SELECT product_cd, SUM(avail_balance) prod_balance
    -> FROM account
    -> WHERE status = 'ACTIVE'
    -> GROUP BY product_cd
    -> HAVING SUM(avail_balance) >= 10000;
+------------+--------------+
| product_cd | prod_balance |
+------------+--------------+
| CD         |     19500.00 |
| CHK        |     73008.01 |
| MM         |     17045.14 |
| SBL        |     50000.00 |
+------------+--------------+
4 rows in set (0.00 sec)
```

This query has two filter conditions: one in the where clause, which filters out nonactive accounts, and the other in the having clause, which filters out any product whose total available balance is less than $10,000. Thus, one of the filters acts on data *before* it is grouped, and the other filter acts on data *after* the groups have been created. If you mistakenly put both filters in the where clause, you will see the following error:

```
mysql> SELECT product_cd, SUM(avail_balance) prod_balance
    -> FROM account
```

```
-> WHERE status = 'ACTIVE'
->    AND SUM(avail_balance) > 10000
-> GROUP BY product_cd;
ERROR 1111 (HY000): Invalid use of group function
```

This query fails because you cannot include an aggregate function in a query's where clause. This is because the filters in the where clause are evaluated *before* the grouping occurs, so the server can't yet perform any functions on groups.

 When adding filters to a query that includes a group by clause, think carefully about whether the filter acts on raw data, in which case it belongs in the where clause, or grouped data, in which case it belongs in the having clause.

You may, however, include aggregate functions in the having clause that do *not* appear in the select clause, as demonstrated by the following:

```
mysql> SELECT product_cd, SUM(avail_balance) prod_balance
    -> FROM account
    -> WHERE status = 'ACTIVE'
    -> GROUP BY product_cd
    -> HAVING MIN(avail_balance) >= 1000
    ->    AND MAX(avail_balance) <= 10000;
+------------+--------------+
| product_cd | prod_balance |
+------------+--------------+
| MM         |     17045.14 |
+------------+--------------+
1 row in set (0.01 sec)
```

This query generates total balances for each product, but the filter condition in the having clause excludes all groups whose minimum balance is less than $1,000 or whose maximum balance is greater than $10,000.

Exercises

Work through the following exercises to test your grasp of SQL's grouping and aggregating features. Check your work with the answers in Appendix C.

8-1

Construct a query that counts the number of rows in the account table.

8-2

Modify your query from exercise 8-1 to count the number of accounts held by each customer. Show the customer ID and the number of accounts for each customer.

8-3

Modify your query from exercise 8-2 to only include those customers having at least two accounts.

8-4 (Extra Credit)

Find the total available balance by product and branch where there is more than one account per product and branch. Order the results by total balance (highest to lowest).

CHAPTER 9

Subqueries

Subqueries are a powerful tool that can be used in all four SQL data statements. This chapter will explore in great detail the many uses of the subquery.

What Is a Subquery?

A *subquery* is a query contained within another SQL statement (which I will refer to as the *containing statement* for the rest of this discussion). A subquery is always enclosed within parentheses, and it is usually executed prior to the containing statement. Like any query, a subquery returns a table that may consist of:

- A single row with a single column
- Multiple rows with a single column
- Multiple rows and columns

The type of table returned by the subquery determines how it may be used and which operators may be used by the containing statement to interact with the table returned by the subquery. When the containing statement has finished executing, the tables returned by any subqueries are discarded, making a subquery act like a temporary table with *statement scope* (meaning that the server frees up any memory allocated to the subquery results after the SQL statement has finished execution).

You have already seen several examples of subqueries in earlier chapters, but here's a simple example to get started:

```
mysql> SELECT account_id, product_cd, cust_id, avail_balance
    -> FROM account
    -> WHERE account_id = (SELECT MAX(account_id) FROM account);
+------------+------------+---------+---------------+
| account_id | product_cd | cust_id | avail_balance |
+------------+------------+---------+---------------+
|         24 | SBL        |      13 |      50000.00 |
+------------+------------+---------+---------------+
1 row in set (0.65 sec)
```

In this example, the subquery returns the maximum value found in the `account_id` column in the `account` table, and the containing statement then returns data about that account. If you are ever confused about what a subquery is doing, you can run the subquery by itself (without the parentheses) to see what it returns. Here's the subquery from the previous example:

```
mysql> SELECT MAX(account_id) FROM account;
+-----------------+
| MAX(account_id) |
+-----------------+
|              24 |
+-----------------+
1 row in set (0.00 sec)
```

So, the subquery returns a single row with a single column, which allows it to be used as one of the expressions in an equality condition (if the subquery returned two or more rows, it could be *compared* to something but could not be *equal* to anything, but more on this later). In this case, you can take the value returned by the subquery and substitute it into the righthand expression of the filter condition in the containing query, as in:

```
mysql> SELECT account_id, product_cd, cust_id, avail_balance
    -> FROM account a
    -> WHERE account_id = 24;
+------------+------------+---------+---------------+
| account_id | product_cd | cust_id | avail_balance |
+------------+------------+---------+---------------+
|         24 | SBL        |      13 |      50000.00 |
+------------+------------+---------+---------------+
1 row in set (0.02 sec)
```

The subquery is useful in this case because it allows you to retrieve information about the highest numbered account in a single query rather than retrieving the maximum `account_id` using one query and then writing a second query to retrieve the desired data from the account table. As you will see, subqueries are useful in many other situations as well and may become one of the most powerful tools in your SQL toolkit.

Subquery Types

Along with the differences noted previously regarding the type of table returned by a subquery (single row/column, single row/multi-column, or multiple columns), there is another factor that can be used to differentiate subqueries; some subqueries are completely self-contained (called *noncorrelated subqueries*), while others reference columns from the containing statement (called *correlated subqueries*). The next several sections will explore these two subquery types and show the different operators that may be employed to interact with them.

Noncorrelated Subqueries

The example from earlier in the chapter is a noncorrelated subquery; it may be executed alone and does not reference anything from the containing statement. Most subqueries that you encounter will be of this type unless you are writing update or delete statements, which frequently make use of correlated subqueries (more on this later). Along with being noncorrelated, the example from earlier in the chapter also returns a table comprised of a single row and column. This type of subquery is known as a *scalar subquery* and can appear on either side of a condition using the usual operators (=, <>, <, >, <=, >=). The next example shows how a scalar subquery may be used in an inequality condition:

```
mysql> SELECT account_id, product_cd, cust_id, avail_balance
    -> FROM account
    -> WHERE open_emp_id <> (SELECT e.emp_id
    ->    FROM employee e INNER JOIN branch b
    ->     ON e.assigned_branch_id = b.branch_id
    ->    WHERE e.title = 'Head Teller' AND b.city = 'Woburn');
+------------+------------+---------+---------------+
| account_id | product_cd | cust_id | avail_balance |
+------------+------------+---------+---------------+
|          6 | CHK        |       3 |       1057.75 |
|          7 | MM         |       3 |       2212.50 |
|          8 | CHK        |       4 |        534.12 |
|          9 | SAV        |       4 |        767.77 |
|         10 | MM         |       4 |       5487.09 |
|         11 | CHK        |       5 |       2237.97 |
|         12 | CHK        |       6 |        122.37 |
|         13 | CD         |       6 |      10000.00 |
|         15 | CHK        |       8 |       3487.19 |
|         16 | SAV        |       8 |        387.99 |
|         17 | CHK        |       9 |        125.67 |
|         18 | MM         |       9 |       9345.55 |
|         19 | CD         |       9 |       1500.00 |
|         20 | CHK        |      10 |      23575.12 |
|         21 | BUS        |      10 |          0.00 |
|         23 | CHK        |      12 |      38552.05 |
|         24 | SBL        |      13 |      50000.00 |
+------------+------------+---------+---------------+
17 rows in set (0.00 sec)
```

This query returns data concerning all accounts that were *not* opened by the head teller at the Woburn branch (the subquery is written using the assumption that there is only a single head teller at each branch). The subquery in this example is a bit more complex than in the previous example, in that it joins two tables and includes two filter conditions. Subqueries may be as simple or as complex as you need them to be, and they may utilize any and all of the available query clauses (select, from, where, group by, having, order by).

If you use a subquery in an equality condition, but the subquery returns more than one row, you will receive an error. For example, if the previous query is modified such that the subquery returns *all* tellers at the Woburn branch instead of the single head teller, you will receive the following error:

```
mysql> SELECT account_id, product_cd, cust_id, avail_balance
    -> FROM account
    -> WHERE open_emp_id <> (SELECT e.emp_id
    ->   FROM employee e INNER JOIN branch b
    ->     ON e.assigned_branch_id = b.branch_id
    ->     WHERE e.title = 'Teller' AND b.city = 'Woburn');
ERROR 1242 (21000): Subquery returns more than 1 row
```

If you run the subquery by itself, you will see the following results:

```
mysql> SELECT e.emp_id
    -> FROM employee e INNER JOIN branch b
    ->   ON e.assigned_branch_id = b.branch_id
    -> WHERE e.title = 'Teller' AND b.city = 'Woburn';
+--------+
| emp_id |
+--------+
|     11 |
|     12 |
+--------+
2 rows in set (0.02 sec)
```

The reason that the containing query fails is that an expression (open_emp_id) cannot be equated to a set of expressions (emp_id's 11 and 12). In other words, a single thing can not be equated to a set of things. In the next section, you will see how to fix the problem by using a different operator.

Multiple-Row, Single-Column Subqueries

If your subquery returns more than one row, then you will not be able to use it on one side of an equality condition, as demonstrated by the previous example. However, there are four additional operators that may be used to build conditions with these types of subqueries.

The in operator

While you can't *equate* a single value to a set of values, you can check to see if a single value can be found *within* a set of values. The next example, while it doesn't use a subquery, demonstrates how to build a condition that uses the in operator to search for a value within a set of values:

```
mysql> SELECT branch_id, name, city
    -> FROM branch
    -> WHERE name IN ('Headquarters', 'Quincy Branch');
+-----------+----------------+---------+
| branch_id | name           | city    |
```

```
+-----------+----------------+---------+
|         1 | Headquarters   | Waltham |
|         3 | Quincy Branch  | Quincy  |
+-----------+----------------+---------+
2 rows in set (0.03 sec)
```

The expression on the lefthand side of the condition is the column name, while the righthand side of the condition is a set of strings. The in operator checks to see if either of the strings can be found in the name column; if so, the condition is met and the row is added to the result set. The same results could be achieved using two equality conditions, as in:

```
mysql> SELECT branch_id, name, city
    -> FROM branch
    -> WHERE name = 'Headquarters' OR name = 'Quincy Branch';
+-----------+----------------+---------+
| branch_id | name           | city    |
+-----------+----------------+---------+
|         1 | Headquarters   | Waltham |
|         3 | Quincy Branch  | Quincy  |
+-----------+----------------+---------+
2 rows in set (0.01 sec)
```

While this approach seems reasonable when the set contains only two expressions, it is easy to see why a single condition using the in operator would be preferable if the set contained dozens (or hundreds, thousands, etc.) of values.

Although you will occasionally create a set of strings, dates, or numbers to use on one side of a condition, you are more likely to generate the set at query execution via a subquery that returns one or more rows. The following query uses the in operator with a subquery on the righthand side of the filter condition to see which employees supervise other employees:

```
mysql> SELECT emp_id, fname, lname, title
    -> FROM employee
    -> WHERE emp_id IN (SELECT superior_emp_id
    ->    FROM employee);
+--------+---------+-----------+--------------------+
| emp_id | fname   | lname     | title              |
+--------+---------+-----------+--------------------+
|      1 | Michael | Smith     | President          |
|      3 | Robert  | Tyler     | Treasurer          |
|      4 | Susan   | Hawthorne | Operations Manager |
|      6 | Helen   | Fleming   | Head Teller        |
|     10 | Paula   | Roberts   | Head Teller        |
|     13 | John    | Blake     | Head Teller        |
|     16 | Theresa | Markham   | Head Teller        |
+--------+---------+-----------+--------------------+
7 rows in set (0.01 sec)
```

The subquery returns the IDs of all employees who supervise other employees, and the containing query retrieves four columns from the employee table for these employees. Here are the results of the subquery:

```
mysql> SELECT superior_emp_id
    -> FROM employee;
+-----------------+
| superior_emp_id |
+-----------------+
|            NULL |
|               1 |
|               1 |
|               3 |
|               4 |
|               4 |
|               4 |
|               4 |
|               4 |
|               6 |
|               6 |
|               6 |
|              10 |
|              10 |
|              13 |
|              13 |
|              16 |
|              16 |
+-----------------+
18 rows in set (0.00 sec)
```

As you can see, some employee IDs are listed more than once, since some employees supervise multiple people. This doesn't adversely affect the results of the containing query, since it doesn't matter whether an employee ID can be found in the result set of the subquery once or more than once. Of course, you could add the distinct keyword to the subquery's select clause if it bothers you to have duplicates in the table returned by the subquery, but it won't change the containing query's result set.

Along with seeing whether a value exists within a set of values, you can check the converse using the not in operator. Here's another version of the previous query using not in instead of in:

```
mysql> SELECT emp_id, fname, lname, title
    -> FROM employee
    -> WHERE emp_id NOT IN (SELECT superior_emp_id
    ->   FROM employee
    ->   WHERE superior_emp_id IS NOT NULL);
+--------+----------+----------+----------------+
| emp_id | fname    | lname    | title          |
+--------+----------+----------+----------------+
|      2 | Susan    | Barker   | Vice President |
|      5 | John     | Gooding  | Loan Manager   |
|      7 | Chris    | Tucker   | Teller         |
|      8 | Sarah    | Parker   | Teller         |
```

```
|     9 | Jane     | Grossman | Teller     |
|    11 | Thomas   | Ziegler  | Teller     |
|    12 | Samantha | Jameson  | Teller     |
|    14 | Cindy    | Mason    | Teller     |
|    15 | Frank    | Portman  | Teller     |
|    17 | Beth     | Fowler   | Teller     |
|    18 | Rick     | Tulman   | Teller     |
+--------+----------+----------+----------------+
11 rows in set (0.00 sec)
```

This query finds all employees who do *not* supervise other people. For this query, I needed to add a filter condition to the subquery to ensure that null values do not appear in the table returned by the subquery; see the next section for an explanation of why this filter is needed in this case.

The all operator

While the in operator is used to see if an expression can be found within a set of expressions, the all operator allows you to make comparisons between a single value and every value in a set. To build such a condition, you will need to use one of the comparison operators (=, <>, <, >, etc.) in conjunction with the all operator. For example, the next query finds all employees whose employee IDs are not equal to any of the supervisor employee IDs:

```
mysql> SELECT emp_id, fname, lname, title
    -> FROM employee
    -> WHERE emp_id <> ALL (SELECT superior_emp_id
    ->    FROM employee
    ->    WHERE superior_emp_id IS NOT NULL);
+--------+----------+----------+----------------+
| emp_id | fname    | lname    | title          |
+--------+----------+----------+----------------+
|      2 | Susan    | Barker   | Vice President |
|      5 | John     | Gooding  | Loan Manager   |
|      7 | Chris    | Tucker   | Teller         |
|      8 | Sarah    | Parker   | Teller         |
|      9 | Jane     | Grossman | Teller         |
|     11 | Thomas   | Ziegler  | Teller         |
|     12 | Samantha | Jameson  | Teller         |
|     14 | Cindy    | Mason    | Teller         |
|     15 | Frank    | Portman  | Teller         |
|     17 | Beth     | Fowler   | Teller         |
|     18 | Rick     | Tulman   | Teller         |
+--------+----------+----------+----------------+
11 rows in set (0.05 sec)
```

Once again, the subquery returns the set of IDs for those employees who supervise other people, and the containing query returns data for each employee whose ID is not equal to all of the IDs returned by the subquery. In other words, the query finds all employees who are not supervisors. If this approach seems a bit clumsy to you, you are in good company; most people would prefer to phrase the query differently and avoid using the all operator. For example, this query generates the same results as the last example in the previous section, which used the not in operator. It's a matter of preference, but I think that most people would find the version that uses not in to be easier to understand.

 When using not in or <> all to compare a value to a set of values, you must be careful to ensure that the set of values does not contain a null value because the server equates the value on the lefthand side of the expression to each member of the set, and any attempt to equate a value to null yields unknown. Thus, the following query returns an empty set:

```
mysql> SELECT emp_id, fname, lname, title
    -> FROM employee
    -> WHERE emp_id NOT IN (1, 2, NULL);
Empty set (0.00 sec)
```

There are cases where the all operator is a bit more natural. The next example uses all to find accounts having an available balance smaller than all of Frank Tucker's accounts:

```
mysql> SELECT account_id, cust_id, product_cd, avail_balance
    -> FROM account
    -> WHERE avail_balance < ALL (SELECT a.avail_balance
    ->   FROM account a INNER JOIN individual i
    ->    ON a.cust_id = i.cust_id
    ->   WHERE i.fname = 'Frank' AND i.lname = 'Tucker');
+------------+---------+------------+---------------+
| account_id | cust_id | product_cd | avail_balance |
+------------+---------+------------+---------------+
|          2 |       1 | SAV        |        500.00 |
|          5 |       2 | SAV        |        200.00 |
|          8 |       4 | CHK        |        534.12 |
|          9 |       4 | SAV        |        767.77 |
|         12 |       6 | CHK        |        122.37 |
|         16 |       8 | SAV        |        387.99 |
|         17 |       9 | CHK        |        125.67 |
|         21 |      10 | BUS        |          0.00 |
+------------+---------+------------+---------------+
8 rows in set (0.01 sec)
```

Here's the table returned by the subquery, which consists of the available balance from each of Frank's accounts:

```
mysql> SELECT a.avail_balance
    -> FROM account a INNER JOIN individual i
    ->   ON a.cust_id = i.cust_id
    -> WHERE i.fname = 'Frank' AND i.lname = 'Tucker';
+---------------+
| avail_balance |
+---------------+
|       1057.75 |
|       2212.50 |
```

```
+----------------+
```
2 rows in set (0.01 sec)

Frank has two accounts, with the lowest balance being $1,057.75. The containing query finds all accounts having a balance smaller than any of Frank's accounts, so the result set includes all accounts having a balance less than $1,057.75.

The any operator

Like the all operator, the any operator allows a value to be compared to the members of a set of values; unlike all, however, a condition using the any operator evaluates to true as soon as a single comparison is favorable, unlike the previous example using the all operator, which only evaluates to true if comparisons against *all* members of the set are favorable. For example, you might want to find all accounts having an available balance greater than *any* of Frank Tucker's accounts:

```
mysql> SELECT account_id, cust_id, product_cd, avail_balance
    -> FROM account
    -> WHERE avail_balance > ANY (SELECT a.avail_balance
    ->   FROM account a INNER JOIN individual i
    ->     ON a.cust_id = i.cust_id
    ->   WHERE i.fname = 'Frank' AND i.lname = 'Tucker');
+------------+---------+------------+---------------+
| account_id | cust_id | product_cd | avail_balance |
+------------+---------+------------+---------------+
|          3 |       1 | CD         |       3000.00 |
|          4 |       2 | CHK        |       2258.02 |
|          7 |       3 | MM         |       2212.50 |
|         10 |       4 | MM         |       5487.09 |
|         11 |       5 | CHK        |       2237.97 |
|         13 |       6 | CD         |      10000.00 |
|         14 |       7 | CD         |       5000.00 |
|         15 |       8 | CHK        |       3487.19 |
|         18 |       9 | MM         |       9345.55 |
|         19 |       9 | CD         |       1500.00 |
|         20 |      10 | CHK        |      23575.12 |
|         22 |      11 | BUS        |       9345.55 |
|         23 |      12 | CHK        |      38552.05 |
|         24 |      13 | SBL        |      50000.00 |
+------------+---------+------------+---------------+
```
14 rows in set (0.01 sec)

Frank has two accounts with balances of $1,057.75 and $2,212.50; to have a balance greater than *any* of these two accounts, an account must have a balance of at least $1,057.75.

While most people prefer to use in, using = any is equivalent to using the in operator.

Multicolumn Subqueries

So far, all of the subquery examples in this chapter have returned a single column and one or more rows. In certain situations, however, you can use subqueries that return two or more columns. To show the utility of multiple-column subqueries, it might help to look first at an example that uses multiple, single-column subqueries:

```
mysql> SELECT account_id, product_cd, cust_id
    -> FROM account
    -> WHERE open_branch_id = (SELECT branch_id
    ->    FROM branch
    ->    WHERE name = 'Woburn Branch')
    ->    AND open_emp_id IN (SELECT emp_id
    ->    FROM employee
    ->    WHERE title = 'Teller' OR title = 'Head Teller');
+------------+------------+---------+
| account_id | product_cd | cust_id |
+------------+------------+---------+
|          1 | CHK        |       1 |
|          2 | SAV        |       1 |
|          3 | CD         |       1 |
|          4 | CHK        |       2 |
|          5 | SAV        |       2 |
|         14 | CD         |       7 |
|         22 | BUS        |      11 |
+------------+------------+---------+
7 rows in set (0.00 sec)
```

This query uses two subqueries to identify the ID of the Woburn branch and the IDs of all bank tellers, and the containing query then uses this information to retrieve all checking accounts opened by a head teller at the Woburn branch. However, since the employee table includes information about which branch each employee is assigned to, you can achieve the same results by comparing both the account.open_branch_id and account.open_emp_id columns to a single subquery against the employee and branch tables. To do so, your filter condition must name both columns from the account table surrounded by parentheses and in the same order as returned by the subquery, as in:

```
mysql> SELECT account_id, product_cd, cust_id
    -> FROM account
    -> WHERE (open_branch_id, open_emp_id) IN
    ->   (SELECT b.branch_id, e.emp_id
    ->    FROM branch b INNER JOIN employee e
    ->      ON b.branch_id = e.assigned_branch_id
    ->    WHERE b.name = 'Woburn Branch'
    ->      AND (e.title = 'Teller' OR e.title = 'Head Teller'));
+------------+------------+---------+
| account_id | product_cd | cust_id |
+------------+------------+---------+
|          1 | CHK        |       1 |
|          2 | SAV        |       1 |
|          3 | CD         |       1 |
```

```
|       4 |  CHK        |        2 |
|       5 |  SAV        |        2 |
|      14 |  CD         |        7 |
|      22 |  BUS        |       11 |
+-----------+------------+---------+
7 rows in set (0.00 sec)
```

This version of the query performs the same function as the previous example, but with a single subquery that returns two columns instead of two subqueries that each return a single column.

Of course, you could rewrite the previous example simply to join the three tables instead of using a subquery, but it's helpful when learning SQL to see multiple ways of achieving the same results. Here's another example, however, that requires a subquery. Let's say that there have been some customer complaints regarding incorrect values in the available/pending balance columns in the account table. Your job is to find all accounts whose balances don't match the sum of the transaction amounts for that account. Here's a partial solution to the problem:

```
SELECT 'ALERT! : Account #1 Has Incorrect Balance!'
FROM account
WHERE (avail_balance, pending_balance) <>
 (SELECT SUM(<expression to generate available balance>),
    SUM(<expression to generate pending balance>)
  FROM transaction
  WHERE account_id = 1)
  AND account_id = 1;
```

As you can see, I have neglected to fill in the expressions used to sum the transaction amounts for the available and pending balance calculations, but I promise to finish the job in Chapter 11 after you learn how to build case expressions. Even so, the query is complete enough to see that the subquery is generating two sums from the transaction table that are then compared to the avail_balance and pending_balance columns in the account table. Both the subquery and the containing query include the filter condition account_id = 1, so the query in its present form will only check a single account at a time. In the next section, you will learn how to write a more general form of the query that will check *all* accounts with a single execution.

Correlated Subqueries

All of the subqueries shown thus far have been independent of their containing statements, meaning that you can execute them by themselves and inspect the results. A *correlated subquery*, on the other hand, is *dependent* on its containing statement from which it references one or more columns. Unlike a noncorrelated subquery, a correlated subquery is not executed once prior to execution of the containing statement; instead, the correlated subquery is executed once for each candidate row (rows that might be included in the final results). For example, the following query uses a correlated subquery to count the number of accounts for each customer, and the containing query then retrieves those customers having exactly two accounts:

```
mysql> SELECT c.cust_id, c.cust_type_cd, c.city
    -> FROM customer c
    -> WHERE 2 = (SELECT COUNT(*)
    ->    FROM account a
    ->    WHERE a.cust_id = c.cust_id);
+---------+--------------+---------+
| cust_id | cust_type_cd | city    |
+---------+--------------+---------+
|       2 | I            | Woburn  |
|       3 | I            | Quincy  |
|       6 | I            | Waltham |
|       8 | I            | Salem   |
|      10 | B            | Salem   |
+---------+--------------+---------+
5 rows in set (0.01 sec)
```

The reference to c.cust_id at the very end of the subquery is what makes the subquery correlated; values for c.cust_id must be supplied by the containing query for the subquery to execute. In this case, the containing query retrieves all thirteen rows from the customer table and executes the subquery once for each customer, passing in the appropriate customer ID for each execution. If the subquery returns the value 2, then the filter condition is met and the row is added to the result set.

Along with equality conditions, you can use correlated subqueries in other types of conditions, such as the range condition illustrated here:

```
mysql> SELECT c.cust_id, c.cust_type_cd, c.city
    -> FROM customer c
    -> WHERE (SELECT SUM(a.avail_balance)
    ->     FROM account a
    ->     WHERE a.cust_id = c.cust_id)
    -> BETWEEN 5000 AND 10000;
+---------+--------------+------------+
| cust_id | cust_type_cd | city       |
+---------+--------------+------------+
|       4 | I            | Waltham    |
|       7 | I            | Wilmington |
|      11 | B            | Wilmington |
+---------+--------------+------------+
3 rows in set (0.02 sec)
```

This variation on the previous query finds all customers whose total available balance across all accounts lies between $5,000 and $10,000. Once again, the correlated subquery is executed 13 times (once for each customer row), and each execution of the subquery returns the total account balance for the given customer.

At the end of the previous section, I demonstrated how to check the available and pending balances of an account against the transactions logged against the account, and I promised to show you how to modify the example to run all accounts in a single execution. Here's the example again:

```
SELECT 'ALERT! : Account #1 Has Incorrect Balance!'
FROM account
WHERE (avail_balance, pending_balance) <>
 (SELECT SUM(<expression to generate available balance>),
    SUM(<expression to generate pending balance>)
  FROM transaction
  WHERE account_id = 1)
 AND account_id = 1;
```

Using a correlated subquery instead of a noncorrelated subquery, you can execute the containing query once, and the subquery will be run for each account. Here's the updated version:

```
SELECT CONCAT('ALERT! : Account #', a.account_id,
  ' Has Incorrect Balance!')
FROM account a
WHERE (a.avail_balance, a.pending_balance) <>
 (SELECT SUM(<expression to generate available balance>),
    SUM(<expression to generate pending balance>)
  FROM transaction t
  WHERE t.account_id = a.account_id);
```

The subquery now includes a filter condition linking the transaction's account ID to the account ID from the containing query. The select clause has also been modified to concatenate an alert message that includes the account ID rather than the hardcoded value 1.

The Exists Operator

While you will often see correlated subqueries used in equality and range conditions, the most common operator used to build conditions that utilize correlated subqueries is the exists operator. The exists operator is used when you want to identify that a relationship exists without regard for the quantity; for example, the following query finds all the accounts for which a transaction was posted on a particular day, without regard for how many transactions were posted:

```
SELECT a.account_id, a.product_cd, a.cust_id, a.avail_balance
FROM account a
WHERE EXISTS (SELECT 1
  FROM transaction t
  WHERE t.account_id = a.account_id
    AND t.txn_date = '2005-01-22');
```

Using the exists operator, your subquery can return zero, one, or many rows, and the condition simply checks whether any rows were returned by the subquery. If you look at the select clause of the subquery, you will see that it consists of a single literal (1); since the condition in the containing query only needs to know how many rows have been returned, the actual data returned by the subquery is irrelevant. Your subquery can return whatever strikes your fancy, as demonstrated next:

```
SELECT a.account_id, a.product_cd, a.cust_id, a.avail_balance
FROM account a
WHERE EXISTS (SELECT t.txn_id, 'hello', 3.1415927
  FROM transaction t
  WHERE t.account_id = a.account_id
    AND t.txn_date = '2005-01-22');
```

However, the convention is to specify either select 1 or select * when using exists.

You may also use not exists to check for subqueries that return no rows, as demonstrated by the following:

```
mysql> SELECT a.account_id, a.product_cd, a.cust_id
    -> FROM account a
    -> WHERE NOT EXISTS (SELECT 1
    ->    FROM business b
    ->    WHERE b.cust_id = a.cust_id);
+------------+------------+---------+
| account_id | product_cd | cust_id |
+------------+------------+---------+
|          1 | CHK        |       1 |
|          2 | SAV        |       1 |
|          3 | CD         |       1 |
|          4 | CHK        |       2 |
|          5 | SAV        |       2 |
|          6 | CHK        |       3 |
|          7 | MM         |       3 |
|          8 | CHK        |       4 |
|          9 | SAV        |       4 |
|         10 | MM         |       4 |
|         11 | CHK        |       5 |
|         12 | CHK        |       6 |
|         13 | CD         |       6 |
|         14 | CD         |       7 |
|         15 | CHK        |       8 |
|         16 | SAV        |       8 |
|         17 | CHK        |       9 |
|         18 | MM         |       9 |
|         19 | CD         |       9 |
+------------+------------+---------+
19 rows in set (0.04 sec)
```

This query finds all customers whose customer ID does not appear in the business table, which is a roundabout way of finding all nonbusiness customers.

Data Manipulation Using Correlated Subqueries

All of the examples thus far in the chapter have been select statements, but don't think that means that subqueries aren't useful in other SQL statements. Subqueries are used heavily in update, delete, and insert statements as well, with correlated subqueries appearing frequently in update and delete statements. Here's an example of a correlated subquery used to modify the last_activity_date column in the account table:

```
UPDATE account a
SET a.last_activity_date =
 (SELECT MAX(t.txn_date)
  FROM transaction t
  WHERE t.account_id = a.account_id);
```

This statement modifies every row in the account table (since there is no where clause) by finding the latest transaction date for each account. While it seems reasonable to expect that every account will have at least one transaction linked to it, it would be best to check whether an account has any transactions before attempting to update the last_activity_date column; otherwise, the column will be set to null, since the subquery would return no rows. Here's another version of the update statement, this time employing a where clause with a second correlated subquery:

```
UPDATE account a
SET a.last_activity_date =
 (SELECT MAX(t.txn_date)
  FROM transaction t
  WHERE t.account_id = a.account_id)
WHERE EXISTS (SELECT 1
  FROM transaction t
  WHERE t.account_id = a.account_id);
```

The two correlated subqueries are identical except for the select clauses. The subquery in the set clause, however, only executes if the condition in the update statement's where clause evaluates to true (meaning that at least one transaction was found for the account), thus protecting the data in the last_activity_date column from being overwritten with a null.

Correlated subqueries are also common in delete statements. For example, you may run a data maintenance script at the end of each month that removes unnecessary data. The script might include the following statement, which removes data from the department table that has no child rows in the employee table:

```
DELETE FROM department
WHERE NOT EXISTS (SELECT 1
  FROM employee
  WHERE employee.dept_id = department.dept_id);
```

When using correlated subqueries with delete statements in MySQL, keep in mind that, for whatever reason, table aliases are not allowed when using delete, which is why I had to use the entire table name in the subquery. With most other database servers, you could provide aliases for the department and employee tables, such as:

```
DELETE FROM department d
WHERE NOT EXISTS (SELECT 1
  FROM employee e
  WHERE e.dept_id = d.dept_id);
```

When to Use Subqueries

Now that you have learned about the different types of subqueries and the different operators that can be employed to interact with the tables returned by subqueries, it's time to explore the many ways in which subqueries can be used to build powerful SQL statements. The next three sections will demonstrate how subqueries may be used to construct custom tables, to build conditions, and to generate column values in result sets.

Subqueries as Data Sources

Back in Chapter 3, I stated that the from clause of a select statement names the *tables* to be used by the query. Since a subquery generates a table containing rows and columns of data, it is perfectly valid to include subqueries in your from clause. Although it might, at first glance, seem like an interesting feature without much practical merit, using subqueries as tables is one of the most powerful tools available when writing queries. Here's a simple example:

```
mysql> SELECT d.dept_id, d.name, e_cnt.how_many num_employees
    -> FROM department d INNER JOIN
    ->  (SELECT dept_id, COUNT(*) how_many
    ->   FROM employee
    ->   GROUP BY dept_id) e_cnt
    ->   ON d.dept_id = e_cnt.dept_id;
+---------+----------------+---------------+
| dept_id | name           | num_employees |
+---------+----------------+---------------+
|       1 | Operations     |            14 |
|       2 | Loans          |             1 |
|       3 | Administration |             3 |
+---------+----------------+---------------+
3 rows in set (0.04 sec)
```

In this example, a subquery generates a list of department IDs along with the number of employees assigned to each department. Here's the table generated by the subquery:

```
mysql> SELECT dept_id, COUNT(*) how_many
    -> FROM employee
    -> GROUP BY dept_id;
+---------+----------+
| dept_id | how_many |
+---------+----------+
|       1 |       14 |
|       2 |        1 |
|       3 |        3 |
+---------+----------+
3 rows in set (0.00 sec)
```

The subquery is given the name e_cnt and is joined to the department table via the dept_id column. The containing query then retrieves the department ID and name from the department table, along with the employee count from the e_cnt subquery.

Subqueries used to generate tables must be noncorrelated; they are executed first, and the tables are held in memory until the containing query finishes execution. Subqueries offer immense flexibility when writing queries, because you can go far beyond the set of available tables to create virtually any view of the data that you desire and then join that table to other tables or subquery-generated tables. If you are writing reports or generating data feeds to external systems, you may be able to do things with a single query that used to demand multiple queries or a procedural language to accomplish.

Table fabrication

Along with using subqueries to summarize existing data, you can use subqueries to generate data that doesn't exist in any form within your database. For example, you may wish to group your customers by the amount of money held in deposit accounts, but you want to use group definitions that are not stored in your database. For example, let's say you want to sort your customers into the following groups:

Group name	Lower limit	Upper limit
Small Fry	0	$4,999.99
Average Joes	$5,000	$9,999.99
Heavy Hitters	$10,000	$9,999,999.99

To generate these groups within a single query, you will need a way to define these three groups. The first step is to define a query that generates the group definitions:

```
mysql> SELECT 'Small Fry' name, 0 low_limit, 4999.99 high_limit
    -> UNION ALL
    -> SELECT 'Average Joes' name, 5000 low_limit, 9999.99 high_limit
    -> UNION ALL
    -> SELECT 'Heavy Hitters' name, 10000 low_limit, 9999999.99 high_limit;
+---------------+-----------+------------+
| name          | low_limit | high_limit |
+---------------+-----------+------------+
| Small Fry     |         0 |    4999.99 |
| Average Joes  |      5000 |    9999.99 |
| Heavy Hitters |     10000 | 9999999.99 |
+---------------+-----------+------------+
3 rows in set (0.00 sec)
```

I have used the set operator union all to merge the results from three separate queries into a single result set. Each query retrieves three literals, and the results from the three queries are put together to generate a table with three rows and three columns.

You now have a query to generate the desired groups, and you can place it into the from clause of another query to generate your customer groups:

```
mysql> SELECT groups.name, COUNT(*) num_customers
    -> FROM
    ->  (SELECT SUM(a.avail_balance) cust_balance
    ->   FROM account a INNER JOIN product p
    ->     ON a.product_cd = p.product_cd
    ->   WHERE p.product_type_cd = 'ACCOUNT'
    ->   GROUP BY a.cust_id) cust_rollup INNER JOIN
    ->  (SELECT 'Small Fry' name, 0 low_limit, 4999.99 high_limit
    ->   UNION ALL
    ->   SELECT 'Average Joes' name, 5000 low_limit,
    ->     9999.99 high_limit
    ->   UNION ALL
    ->   SELECT 'Heavy Hitters' name, 10000 low_limit,
    ->     9999999.99 high_limit) groups
    ->   ON cust_rollup.cust_balance
    ->     BETWEEN groups.low_limit AND groups.high_limit
    -> GROUP BY groups.name;
+---------------+---------------+
| name          | num_customers |
+---------------+---------------+
| Average Joes  |             2 |
| Heavy Hitters |             4 |
| Small Fry     |             5 |
+---------------+---------------+
3 rows in set (0.01 sec)
```

There are two subqueries in the from clause; the first subquery, named cust_rollup, returns the total deposit balances for each customer, while the second subquery, named groups, generates the table containing the three customer groupings. Here's the table generated by cust_rollup:

```
mysql> SELECT SUM(a.avail_balance) cust_balance
    -> FROM account a INNER JOIN product p
    ->   ON a.product_cd = p.product_cd
    -> WHERE p.product_type_cd = 'ACCOUNT'
    -> GROUP BY a.cust_id;
+--------------+
| cust_balance |
+--------------+
|      4557.75 |
|      2458.02 |
|      3270.25 |
|      6788.98 |
|      2237.97 |
|     10122.37 |
|      5000.00 |
|      3875.18 |
|     10971.22 |
|     23575.12 |
|     38552.05 |
+--------------+
11 rows in set (0.05 sec)
```

The table generated by cust_rollup is then joined to the groups table via a range condition (cust_rollup.cust_balance BETWEEN groups.low_limit AND groups.high_limit). Finally, the joined data is grouped and the number of customers in each group is counted to generate the final result set.

Of course, you could simply decide to build a permanent table to hold the group definitions instead of using a subquery. Using that approach, you would find your database to be littered with small special-purpose tables after a while, and you wouldn't remember the reason for which most of them were created. I've worked in environments where the database users were allowed to create their own tables for special purposes, and the results were disastrous (tables not included in backups, tables lost during server upgrades, server downtime due to space allocation issues, etc.). Armed with subqueries, however, you will be able to adhere to a policy where tables are only added to a database when there is a clear business need to store new data.

Task-oriented subqueries

In systems used for reporting or data-feed generation, you will often come across queries such as the following:

```
mysql> SELECT p.name product, b.name branch,
    ->   CONCAT(e.fname, ' ', e.lname) name,
    ->   SUM(a.avail_balance) tot_deposits
    -> FROM account a INNER JOIN employee e
    ->   ON a.open_emp_id = e.emp_id
    ->   INNER JOIN branch b
    ->   ON a.open_branch_id = b.branch_id
    ->   INNER JOIN product p
    ->   ON a.product_cd = p.product_cd
    -> WHERE p.product_type_cd = 'ACCOUNT'
    -> GROUP BY p.name, b.name, e.fname, e.lname;
+-----------------------+---------------+-----------------+--------------+
| product               | branch        | name            | tot_deposits |
+-----------------------+---------------+-----------------+--------------+
| certificate of deposit | Headquarters | Michael Smith   |     11500.00 |
| certificate of deposit | Woburn Branch | Paula Roberts   |      8000.00 |
| checking account      | Headquarters  | Michael Smith   |       782.16 |
| checking account      | Quincy Branch | John Blake      |      1057.75 |
| checking account      | So. NH Branch | Theresa Markham |     67852.33 |
| checking account      | Woburn Branch | Paula Roberts   |      3315.77 |
| money market account  | Headquarters  | Michael Smith   |     14832.64 |
| money market account  | Quincy Branch | John Blake      |      2212.50 |
| savings account       | Headquarters  | Michael Smith   |       767.77 |
| savings account       | So. NH Branch | Theresa Markham |       387.99 |
| savings account       | Woburn Branch | Paula Roberts   |       700.00 |
+-----------------------+---------------+-----------------+--------------+
11 rows in set (0.02 sec)
```

This query sums all deposit account balances by account type, the employee that opened the accounts, and the branches at which the accounts were opened. If you look at the query closely, you will see that the product, branch, and employee tables are only needed for display purposes, and that the account table has everything needed to generate the groupings (product_cd, open_branch_id, open_emp_id, and avail_balance). Therefore, you could separate out the task of generating the groups into a subquery, and then join the other three tables to the table generated by the subquery to achieve the desired end result. Here's the grouping subquery:

```
mysql> SELECT product_cd, open_branch_id branch_id, open_emp_id emp_id,
    ->   SUM(avail_balance) tot_deposits
    -> FROM account
    -> GROUP BY product_cd, open_branch_id, open_emp_id;
+------------+-----------+--------+--------------+
| product_cd | branch_id | emp_id | tot_deposits |
+------------+-----------+--------+--------------+
| BUS        |         2 |     10 |      9345.55 |
| BUS        |         4 |     16 |         0.00 |
| CD         |         1 |      1 |     11500.00 |
| CD         |         2 |     10 |      8000.00 |
| CHK        |         1 |      1 |       782.16 |
| CHK        |         2 |     10 |      3315.77 |
| CHK        |         3 |     13 |      1057.75 |
| CHK        |         4 |     16 |     67852.33 |
| MM         |         1 |      1 |     14832.64 |
| MM         |         3 |     13 |      2212.50 |
| SAV        |         1 |      1 |       767.77 |
| SAV        |         2 |     10 |       700.00 |
| SAV        |         4 |     16 |       387.99 |
| SBL        |         3 |     13 |     50000.00 |
+------------+-----------+--------+--------------+
14 rows in set (0.01 sec)
```

This is the heart of the query; the other tables are only needed to provide meaningful strings in place of the product_cd, open_branch_id, and open_emp_id foreign-key columns. The next query wraps the query against the account table in a subquery and joins the resulting table to the other three tables:

```
mysql> SELECT p.name product, b.name branch,
    ->   CONCAT(e.fname, ' ', e.lname) name,
    ->   account_groups.tot_deposits
    -> FROM
    ->  (SELECT product_cd, open_branch_id branch_id,
    ->     open_emp_id emp_id,
    ->     SUM(avail_balance) tot_deposits
    ->   FROM account
    ->   GROUP BY product_cd, open_branch_id, open_emp_id) account_groups
    ->   INNER JOIN employee e ON e.emp_id = account_groups.emp_id
    ->   INNER JOIN branch b ON b.branch_id = account_groups.branch_id
    ->   INNER JOIN product p ON p.product_cd = account_groups.product_cd
    -> WHERE p.product_type_cd = 'ACCOUNT';
+-----------------------+--------------+---------------+--------------+
| product               | branch       | name          | tot_deposits |
+-----------------------+--------------+---------------+--------------+
| certificate of deposit | Headquarters | Michael Smith |     11500.00 |
| checking account      | Headquarters | Michael Smith |       782.16 |
```

```
| money market account  | Headquarters  | Michael Smith   |    14832.64 |
| savings account       | Headquarters  | Michael Smith   |      767.77 |
| certificate of deposit| Woburn Branch | Paula Roberts   |     8000.00 |
| checking account      | Woburn Branch | Paula Roberts   |     3315.77 |
| savings account       | Woburn Branch | Paula Roberts   |      700.00 |
| checking account      | Quincy Branch | John Blake      |     1057.75 |
| money market account  | Quincy Branch | John Blake      |     2212.50 |
| checking account      | So. NH Branch | Theresa Markham |    67852.33 |
| savings account       | So. NH Branch | Theresa Markham |      387.99 |
+-----------------------+---------------+-----------------+-------------+
11 rows in set (0.00 sec)
```

I realize that beauty is in the eyes of the beholder, but I find this version of the query to be far more satisfying than the big, flat version. This version may execute faster, as well, because the grouping is being done on small foreign-key columns (product_cd, open_branch_id, open_emp_id) instead of potentially lengthy string columns (branch.name, product.name, employee.fname, employee.lname).

Subqueries in Filter Conditions

Many of the examples in this chapter have used subqueries as expressions in filter conditions, so it should not surprise you that this is one of the main uses for subqueries. However, filter conditions using subqueries are not found only in the where clause. For example, the next query uses a subquery in the having clause to find the employee responsible for opening the most accounts:

```
mysql> SELECT open_emp_id, COUNT(*) how_many
    -> FROM account
    -> GROUP BY open_emp_id
    -> HAVING COUNT(*) = (SELECT MAX(emp_cnt.how_many)
    ->   FROM (SELECT COUNT(*) how_many
    ->     FROM account
    ->     GROUP BY open_emp_id) emp_cnt);
+-------------+----------+
| open_emp_id | how_many |
+-------------+----------+
|           1 |        8 |
+-------------+----------+
1 row in set (0.01 sec)
```

The subquery in the having clause finds the maximum number of accounts opened by any employee, and the containing query finds the employee that has opened that number of accounts. If multiple employees tie for the highest number of opened accounts, then the query would return multiple rows.

Subqueries as Expression Generators

For this last section of the chapter, I will finish where I began: with single-column, single-row scalar subqueries. Along with being used in filter conditions, scalar subqueries may be used wherever an expression can appear, including the select and order by clauses of a query and the values clause of an insert statement.

Earlier in the chapter, in the section entitled "Task-oriented subqueries," I showed how to use a subquery to separate out the grouping mechanism from the rest of the query. Here's another version of the same query that uses subqueries for the same purpose, but in a different way:

```
mysql> SELECT
    ->   (SELECT p.name FROM product p
    ->    WHERE p.product_cd = a.product_cd
    ->      AND p.product_type_cd = 'ACCOUNT') product,
    ->   (SELECT b.name FROM branch b
    ->    WHERE b.branch_id = a.open_branch_id) branch,
    ->   (SELECT CONCAT(e.fname, ' ', e.lname) FROM employee e
    ->    WHERE e.emp_id = a.open_emp_id) name,
    ->   SUM(a.avail_balance) tot_deposits
    -> FROM account a
    -> GROUP BY a.product_cd, a.open_branch_id, a.open_emp_id;
+-----------------------+----------------+------------------+--------------+
| product               | branch         | name             | tot_deposits |
+-----------------------+----------------+------------------+--------------+
| NULL                  | Woburn Branch  | Paula Roberts    |      9345.55 |
| NULL                  | So. NH Branch  | Theresa Markham  |         0.00 |
| certificate of deposit| Headquarters   | Michael Smith    |     11500.00 |
| certificate of deposit| Woburn Branch  | Paula Roberts    |      8000.00 |
| checking account      | Headquarters   | Michael Smith    |       782.16 |
| checking account      | Woburn Branch  | Paula Roberts    |      3315.77 |
| checking account      | Quincy Branch  | John Blake       |      1057.75 |
| checking account      | So. NH Branch  | Theresa Markham  |     67852.33 |
| money market account  | Headquarters   | Michael Smith    |     14832.64 |
| money market account  | Quincy Branch  | John Blake       |      2212.50 |
| savings account       | Headquarters   | Michael Smith    |       767.77 |
| savings account       | Woburn Branch  | Paula Roberts    |       700.00 |
| savings account       | So. NH Branch  | Theresa Markham  |       387.99 |
| NULL                  | Quincy Branch  | John Blake       |     50000.00 |
+-----------------------+----------------+------------------+--------------+
14 rows in set (0.01 sec)
```

There are two main differences between this query and the earlier version using a subquery in the from clause:

- Instead of joining the product, branch, and employee tables to the account data, correlated scalar subqueries are used in the select clause to look up the product, branch, and employee names.

- The result set has 14 rows instead of 11 rows, and three of the product names are null.

The reason for the extra three rows in the result set is that the previous version of the query included the filter condition p.product_type_cd = 'ACCOUNT'. That filter eliminated rows with product types of "INSURANCE" and "LOAN," such as small business loans. Since this version of the query doesn't include a join to the product table, there is no way to include the filter condition in the main query. The correlated subquery against the product table does include this filter, but the only effect is to leave the product name null. If you want to get rid of the extra three rows, you could join the product table to the account table and include the filter condition, or you could simply do the following:

```
mysql> SELECT all_prods.product, all_prods.branch,
    ->   all_prods.name, all_prods.tot_deposits
    -> FROM
    -> (SELECT
    ->   (SELECT p.name FROM product p
    ->    WHERE p.product_cd = a.product_cd
    ->      AND p.product_type_cd = 'ACCOUNT') product,
    ->   (SELECT b.name FROM branch b
    ->    WHERE b.branch_id = a.open_branch_id) branch,
    ->   (SELECT CONCAT(e.fname, ' ', e.lname) FROM employee e
    ->    WHERE e.emp_id = a.open_emp_id) name,
    ->   SUM(a.avail_balance) tot_deposits
    ->  FROM account a
    ->  GROUP BY a.product_cd, a.open_branch_id, a.open_emp_id) all_prods
    -> WHERE all_prods.product IS NOT NULL;
+-------------------------+----------------+-----------------+--------------+
| product                 | branch         | name            | tot_deposits |
+-------------------------+----------------+-----------------+--------------+
| certificate of deposit  | Headquarters   | Michael Smith   |     11500.00 |
| certificate of deposit  | Woburn Branch  | Paula Roberts   |      8000.00 |
| checking account        | Headquarters   | Michael Smith   |       782.16 |
| checking account        | Woburn Branch  | Paula Roberts   |      3315.77 |
| checking account        | Quincy Branch  | John Blake      |      1057.75 |
| checking account        | So. NH Branch  | Theresa Markham |     67852.33 |
| money market account    | Headquarters   | Michael Smith   |     14832.64 |
| money market account    | Quincy Branch  | John Blake      |      2212.50 |
| savings account         | Headquarters   | Michael Smith   |       767.77 |
| savings account         | Woburn Branch  | Paula Roberts   |       700.00 |
| savings account         | So. NH Branch  | Theresa Markham |       387.99 |
+-------------------------+----------------+-----------------+--------------+
11 rows in set (0.01 sec)
```

Simply by wrapping the previous query in a subquery (called all_prods) and adding a filter condition to exclude null values of the product column, the query now returns the desired 11 rows. The end result is a query that performs all grouping against raw data in the account table, and then embellishes the output using data in three other tables, and *without doing any joins*.

As previously noted, scalar subqueries can also appear in the order by clause. The following query retrieves employee data sorted by the last name of each employee's boss, and then by the employee's last name:

```
mysql> SELECT emp.emp_id, CONCAT(emp.fname, ' ', emp.lname) emp_name,
    ->   (SELECT CONCAT(boss.fname, ' ', boss.lname)
    ->    FROM employee boss
    ->    WHERE boss.emp_id = emp.superior_emp_id) boss_name
```

```
  -> FROM employee emp
  -> WHERE emp.superior_emp_id IS NOT NULL
  -> ORDER BY (SELECT boss.lname FROM employee boss
  ->   WHERE boss.emp_id = emp.superior_emp_id), emp.lname;
+---------+------------------+------------------+
| emp_id  | emp_name         | boss_name        |
+---------+------------------+------------------+
|      14 | Cindy Mason      | John Blake       |
|      15 | Frank Portman    | John Blake       |
|       9 | Jane Grossman    | Helen Fleming    |
|       8 | Sarah Parker     | Helen Fleming    |
|       7 | Chris Tucker     | Helen Fleming    |
|      13 | John Blake       | Susan Hawthorne  |
|       6 | Helen Fleming    | Susan Hawthorne  |
|       5 | John Gooding     | Susan Hawthorne  |
|      16 | Theresa Markham  | Susan Hawthorne  |
|      10 | Paula Roberts    | Susan Hawthorne  |
|      17 | Beth Fowler      | Theresa Markham  |
|      18 | Rick Tulman      | Theresa Markham  |
|      12 | Samantha Jameson | Paula Roberts    |
|      11 | Thomas Ziegler   | Paula Roberts    |
|       2 | Susan Barker     | Michael Smith    |
|       3 | Robert Tyler     | Michael Smith    |
|       4 | Susan Hawthorne  | Robert Tyler     |
+---------+------------------+------------------+
17 rows in set (0.01 sec)
```

The query uses two correlated scalar subqueries: one in the select clause to retrieve the full name of each employee's boss, and another in the order by clause to return just the last name of each employee's boss for sorting purposes.

Along with using correlated scalar subqueries in select statements, you can use non-correlated scalar subqueries to generate values for an insert statement. For example, let's say you are going to generate a new account row, and you've been given the following data:

- The product name ("savings account")
- The customer's Federal ID ("555-55-5555")
- The name of the branch where the account was opened ("Quincy Branch")
- The first and last names of the teller who opened the account ("Frank Portman")

Before you can create a row in the account table, you will need to look up the key values for all of these pieces of data so that you can populate the foreign-key columns in the account table. You have two choices for how to go about it: execute four queries to retrieve the primary-key values and place those values into an insert statement, or use subqueries to retrieve the four key values from within an insert statement. Here's an example of the latter approach:

```
INSERT INTO account
  (account_id, product_cd, cust_id, open_date, last_activity_date,
   status, open_branch_id, open_emp_id, avail_balance, pending_balance)
```

```
VALUES (NULL,
 (SELECT product_cd FROM product WHERE name = 'savings account'),
 (SELECT cust_id FROM customer WHERE fed_id = '555-55-5555'),
  '2005-01-25', '2005-01-25', 'ACTIVE',
 (SELECT branch_id FROM branch WHERE name = 'Quincy Branch'),
 (SELECT emp_id FROM employee WHERE lname = 'Portman' AND fname = 'Frank'),
  0, 0);
```

Using a single SQL statement, you can create a row in the account table and look up four foreign-key column values at the same time. There is one downside to this approach, however. When using subqueries to generate data for columns that allow null values, your insert statement will succeed even if one of your subqueries fails to return a value. For example, if you mistyped Frank Portman's name in the fourth subquery, a row will still be created in account, but the open_emp_id would be set to null.

Subquery Wrap-up

I have covered a lot of ground in this chapter, so it might be a good idea to review it. The examples used in this chapter have demonstrated subqueries that:

- Return a single column and row, a single column with multiple rows, and multiple columns and rows
- Are independent of the containing statement (noncorrelated subqueries)
- Reference one or more columns from the containing statement (correlated subqueries)
- Are used in conditions that utilize comparison operators as well as the special-purpose operators in, not in, exists, and not exists
- Can be found in select, update, delete, and insert statements
- Generate tables that can be joined to other tables in a query
- Can be used to generate values to populate a table or to populate columns in a query's result set
- Are used in the select, from, where, having, and order by clauses of queries

Obviously, subqueries are a very versatile tool, so don't feel bad if all of these concepts haven't sunk in after reading this chapter for the first time. Keep experimenting with the various uses for subqueries, and you will soon find yourself thinking about how you might utilize a subquery every time you write a nontrivial SQL statement.

Exercises

These exercises are designed to test your understanding of subqueries. Please see Appendix C for solutions.

9-1

Construct a query against the account table that uses a filter condition with a noncorrelated subquery against the product table to find all loan accounts (product. product_type_cd = 'LOAN'). Retrieve the account ID, product code, customer ID, and available balance.

9-2

Rework the query from exercise 9-1 using a *correlated* subquery against the product table to achieve the same results.

9-3

Join the following query to the employee table to show the experience level of each employee:

```
SELECT 'trainee' name, '2004-01-01' start_dt, '2005-12-31' end_dt
UNION ALL
SELECT 'worker' name, '2002-01-01' start_dt, '2003-12-31' end_dt
UNION ALL
SELECT 'mentor' name, '2000-01-01' start_dt, '2001-12-31' end_dt
```

Give the subquery the alias "levels," and include the employee ID, first name, last name, and experience level (levels.name). (Hint: Build a join condition using an inequality condition to determine into which level the employee.start_date column falls.)

9-4

Construct a query against the employee table that retrieves the employee ID, first name, and last name, along with the names of the department and branch to which the employee is assigned. Do not join any tables.

CHAPTER 10
Joins Revisited

By now, you should be comfortable with the concept of the inner join, which was introduced in Chapter 5. This chapter will focus on other ways in which tables can be joined, including the outer join and the cross join.

Outer Joins

In all of the examples thus far that have included multiple tables, there hasn't been any concern that the join conditions might fail to find matches for all of the rows in the tables. For example, when joining the account table to the customer table, no mention was made of the possibility that a value in the cust_id column of the account table might not match a value in the cust_id column of the customer table. If that were the case, then some of the rows in one table or the other would be left out of the result set.

Just to be sure, let's check the data in the tables. Here are the account_id and cust_id columns from the account table:

```
mysql> SELECT account_id, cust_id
    -> FROM account;
+------------+---------+
| account_id | cust_id |
+------------+---------+
|          1 |       1 |
|          2 |       1 |
|          3 |       1 |
|          4 |       2 |
|          5 |       2 |
|          6 |       3 |
|          7 |       3 |
|          8 |       4 |
|          9 |       4 |
|         10 |       4 |
|         11 |       5 |
|         12 |       6 |
```

```
|    13 |      6 |
|    14 |      7 |
|    15 |      8 |
|    16 |      8 |
|    17 |      9 |
|    18 |      9 |
|    19 |      9 |
|    20 |     10 |
|    21 |     10 |
|    22 |     11 |
|    23 |     12 |
|    24 |     13 |
+-----------+---------+
24 rows in set (0.04 sec)
```

There are 24 accounts spanning 13 different customers, with customer IDs 1 through 13 having at least one account. Here's the set of customer IDs from the customer table:

```
mysql> SELECT cust_id
    -> FROM customer;

+---------+
| cust_id |
+---------+
|       1 |
|       2 |
|       3 |
|       4 |
|       5 |
|       6 |
|       7 |
|       8 |
|       9 |
|      10 |
|      11 |
|      12 |
|      13 |
+---------+
13 rows in set (0.02 sec)
```

There are 13 rows in the customer table with IDs 1 through 13, so every customer ID is included at least once in the account table. When the two tables are joined on the cust_id column, therefore, you would expect all 24 rows to be included in the result set (barring any other filter conditions):

```
mysql> SELECT a.account_id, c.cust_id
    -> FROM account a INNER JOIN customer c
    ->   ON a.cust_id = c.cust_id;

+------------+---------+
| account_id | cust_id |
+------------+---------+
|          1 |       1 |
```

```
|           2 |       1 |
|           3 |       1 |
|           4 |       2 |
|           5 |       2 |
|           6 |       3 |
|           7 |       3 |
|           8 |       4 |
|           9 |       4 |
|          10 |       4 |
|          11 |       5 |
|          12 |       6 |
|          13 |       6 |
|          14 |       7 |
|          15 |       8 |
|          16 |       8 |
|          17 |       9 |
|          18 |       9 |
|          19 |       9 |
|          20 |      10 |
|          21 |      10 |
|          22 |      11 |
|          23 |      12 |
|          24 |      13 |
+-------------+---------+
24 rows in set (0.00 sec)
```

As expected, all 24 accounts are present in the result set. But what happens if you join the account table to one of the specialized customer tables, such as the business table?

```
mysql> SELECT a.account_id, b.cust_id, b.name
    -> FROM account a INNER JOIN business b
    ->   ON a.cust_id = b.cust_id;
+------------+---------+-----------------------+
| account_id | cust_id | name                  |
+------------+---------+-----------------------+
|         20 |      10 | Chilton Engineering   |
|         21 |      10 | Chilton Engineering   |
|         22 |      11 | Northeast Cooling Inc.|
|         23 |      12 | Superior Auto Body    |
|         24 |      13 | AAA Insurance Inc.    |
+------------+---------+-----------------------+
5 rows in set (0.00 sec)
```

Instead of 24 rows in the result set, there are now only five. Let's look in the business table to see why this is:

```
mysql> SELECT cust_id, name
    -> FROM business;
+---------+-----------------------+
| cust_id | name                  |
+---------+-----------------------+
|      10 | Chilton Engineering   |
|      11 | Northeast Cooling Inc.|
```

```
|      12 | Superior Auto Body   |
|      13 | AAA Insurance Inc.   |
+---------+---------------------+
4 rows in set (0.01 sec)
```

Of the 13 rows in the customer table, only four are business customers, and since one of the business customers has two accounts, there are a total of five rows in the account table that are linked to business customers.

But what if you want your query to return *all* of the accounts, but to include the business name only if the account is linked to a business customer? This is an example where you would need an *outer join* between the account and business tables, as in:

```
mysql> SELECT a.account_id, a.cust_id, b.name
    -> FROM account a LEFT OUTER JOIN business b
    ->   ON a.cust_id = b.cust_id;
+------------+---------+------------------------+
| account_id | cust_id | name                   |
+------------+---------+------------------------+
|          1 |       1 | NULL                   |
|          2 |       1 | NULL                   |
|          3 |       1 | NULL                   |
|          4 |       2 | NULL                   |
|          5 |       2 | NULL                   |
|          6 |       3 | NULL                   |
|          7 |       3 | NULL                   |
|          8 |       4 | NULL                   |
|          9 |       4 | NULL                   |
|         10 |       4 | NULL                   |
|         11 |       5 | NULL                   |
|         12 |       6 | NULL                   |
|         13 |       6 | NULL                   |
|         14 |       7 | NULL                   |
|         15 |       8 | NULL                   |
|         16 |       8 | NULL                   |
|         17 |       9 | NULL                   |
|         18 |       9 | NULL                   |
|         19 |       9 | NULL                   |
|         20 |      10 | Chilton Engineering    |
|         21 |      10 | Chilton Engineering    |
|         22 |      11 | Northeast Cooling Inc. |
|         23 |      12 | Superior Auto Body     |
|         24 |      13 | AAA Insurance Inc.     |
+------------+---------+------------------------+
24 rows in set (0.00 sec)
```

An outer join includes all of the rows from one table and includes data from the second table only if matching rows are found. In this case, all rows from the account table are included, since I specified left outer join and the account table is on the left side of the join definition. The name column is null for all rows except for the four business customers (cust_id 10, 11, 12, and 13). Here's a similar query with an outer join to the individual table instead of the business table:

```
mysql> SELECT a.account_id, a.cust_id, i.fname, i.lname
    -> FROM account a LEFT OUTER JOIN individual i
    ->   ON a.cust_id = i.cust_id;
+------------+---------+----------+---------+
| account_id | cust_id | fname    | lname   |
+------------+---------+----------+---------+
|          1 |       1 | James    | Hadley  |
|          2 |       1 | James    | Hadley  |
|          3 |       1 | James    | Hadley  |
|          4 |       2 | Susan    | Tingley |
|          5 |       2 | Susan    | Tingley |
|          6 |       3 | Frank    | Tucker  |
|          7 |       3 | Frank    | Tucker  |
|          8 |       4 | John     | Hayward |
|          9 |       4 | John     | Hayward |
|         10 |       4 | John     | Hayward |
|         11 |       5 | Charles  | Frasier |
|         12 |       6 | John     | Spencer |
|         13 |       6 | John     | Spencer |
|         14 |       7 | Margaret | Young   |
|         15 |       8 | Louis    | Blake   |
|         16 |       8 | Louis    | Blake   |
|         17 |       9 | Richard  | Farley  |
|         18 |       9 | Richard  | Farley  |
|         19 |       9 | Richard  | Farley  |
|         20 |      10 | NULL     | NULL    |
|         21 |      10 | NULL     | NULL    |
|         22 |      11 | NULL     | NULL    |
|         23 |      12 | NULL     | NULL    |
|         24 |      13 | NULL     | NULL    |
+------------+---------+----------+---------+
24 rows in set (0.00 sec)
```

This query is essentially the reverse of the previous query: first and last names are supplied for the individual customers, whereas the columns are null for the business customers.

Left Versus Right Outer Joins

In the outer join examples in the previous section, I specified left outer join. The keyword left indicates that the table on the left side of the from clause is responsible for determining the number of rows in the result set, whereas the table on the right side is used to provide column values whenever a match is found. Consider the following query:

```
mysql> SELECT c.cust_id, b.name
    -> FROM customer c LEFT OUTER JOIN business b
    ->   ON c.cust_id = b.cust_id;
```

```
+---------+-----------------------+
| cust_id | name                  |
+---------+-----------------------+
|       1 | NULL                  |
|       2 | NULL                  |
|       3 | NULL                  |
|       4 | NULL                  |
|       5 | NULL                  |
|       6 | NULL                  |
|       7 | NULL                  |
|       8 | NULL                  |
|       9 | NULL                  |
|      10 | Chilton Engineering   |
|      11 | Northeast Cooling Inc. |
|      12 | Superior Auto Body    |
|      13 | AAA Insurance Inc.    |
+---------+-----------------------+
13 rows in set (0.00 sec)
```

The from clause specifies a left outer join, so all 13 rows from the customer table are included in the result set, with the business table contributing values to the second column in the result set for the four business customers. If you execute the same query, but indicate a right outer join, you would see the following results:

```
mysql> SELECT c.cust_id, b.name
    -> FROM customer c RIGHT OUTER JOIN business b
    ->   ON c.cust_id = b.cust_id;
+---------+-----------------------+
| cust_id | name                  |
+---------+-----------------------+
|      10 | Chilton Engineering   |
|      11 | Northeast Cooling Inc. |
|      12 | Superior Auto Body    |
|      13 | AAA Insurance Inc.    |
+---------+-----------------------+
4 rows in set (0.00 sec)
```

The number of rows in the result set is now determined by the number of rows in the business table, which is why there are only four rows in the result set.

Keep in mind that both queries are performing outer joins; the keywords left and right are there just to tell the database optimizer which table is allowed to have gaps in the data. If you want to outer join tables A and B and you want all rows from A with rows from B whenever there is matching data, you can specify either A left outer join B or B right outer join A.

Three-Way Outer Joins

In some cases, you may want to outer join one table with two other tables. For example, you may want a list of all accounts showing either the customer's first and last names for individuals or the business name for business customers, as in:

```
mysql> SELECT a.account_id, a.product_cd,
    ->   CONCAT(i.fname, ' ', i.lname) person_name,
    ->   b.name business_name
    -> FROM account a LEFT OUTER JOIN individual i
    ->   ON a.cust_id = i.cust_id
    ->   LEFT OUTER JOIN business b
    ->   ON a.cust_id = b.cust_id;
+------------+------------+-----------------+----------------------+
| account_id | product_cd | person_name     | business_name        |
+------------+------------+-----------------+----------------------+
|          1 | CHK        | James Hadley    | NULL                 |
|          2 | SAV        | James Hadley    | NULL                 |
|          3 | CD         | James Hadley    | NULL                 |
|          4 | CHK        | Susan Tingley   | NULL                 |
|          5 | SAV        | Susan Tingley   | NULL                 |
|          6 | CHK        | Frank Tucker    | NULL                 |
|          7 | MM         | Frank Tucker    | NULL                 |
|          8 | CHK        | John Hayward    | NULL                 |
|          9 | SAV        | John Hayward    | NULL                 |
|         10 | MM         | John Hayward    | NULL                 |
|         11 | CHK        | Charles Frasier | NULL                 |
|         12 | CHK        | John Spencer    | NULL                 |
|         13 | CD         | John Spencer    | NULL                 |
|         14 | CD         | Margaret Young  | NULL                 |
|         15 | CHK        | Louis Blake     | NULL                 |
|         16 | SAV        | Louis Blake     | NULL                 |
|         17 | CHK        | Richard Farley  | NULL                 |
|         18 | MM         | Richard Farley  | NULL                 |
|         19 | CD         | Richard Farley  | NULL                 |
|         20 | CHK        | NULL            | Chilton Engineering  |
|         21 | BUS        | NULL            | Chilton Engineering  |
|         22 | BUS        | NULL            | Northeast Cooling Inc. |
|         23 | CHK        | NULL            | Superior Auto Body   |
|         24 | SBL        | NULL            | AAA Insurance Inc.   |
+------------+------------+-----------------+----------------------+
24 rows in set (0.00 sec)
```

The results include all 24 rows from the account table, along with either a person's name or a business name coming from the two outer-joined tables.

I don't know of any restrictions with MySQL regarding the number of tables that can be outer-joined to the same table, but you can always use subqueries to limit the number of joins in your query. For example, the previous example can be rewritten as follows:

```
mysql> SELECT account_ind.account_id, account_ind.product_cd,
    ->   account_ind.person_name,
```

```
    ->    b.name business_name
    -> FROM
    ->    (SELECT a.account_id, a.product_cd, a.cust_id,
    ->        CONCAT(i.fname, ' ', i.lname) person_name
    ->     FROM account a LEFT OUTER JOIN individual i
    ->       ON a.cust_id = i.cust_id) account_ind
    ->     LEFT OUTER JOIN business b
    ->       ON account_ind.cust_id = b.cust_id;
+------------+------------+------------------+------------------------+
| account_id | product_cd | person_name      | business_name          |
+------------+------------+------------------+------------------------+
|          1 | CHK        | James Hadley     | NULL                   |
|          2 | SAV        | James Hadley     | NULL                   |
|          3 | CD         | James Hadley     | NULL                   |
|          4 | CHK        | Susan Tingley    | NULL                   |
|          5 | SAV        | Susan Tingley    | NULL                   |
|          6 | CHK        | Frank Tucker     | NULL                   |
|          7 | MM         | Frank Tucker     | NULL                   |
|          8 | CHK        | John Hayward     | NULL                   |
|          9 | SAV        | John Hayward     | NULL                   |
|         10 | MM         | John Hayward     | NULL                   |
|         11 | CHK        | Charles Frasier  | NULL                   |
|         12 | CHK        | John Spencer     | NULL                   |
|         13 | CD         | John Spencer     | NULL                   |
|         14 | CD         | Margaret Young   | NULL                   |
|         15 | CHK        | Louis Blake      | NULL                   |
|         16 | SAV        | Louis Blake      | NULL                   |
|         17 | CHK        | Richard Farley   | NULL                   |
|         18 | MM         | Richard Farley   | NULL                   |
|         19 | CD         | Richard Farley   | NULL                   |
|         20 | CHK        | NULL             | Chilton Engineering    |
|         21 | BUS        | NULL             | Chilton Engineering    |
|         22 | BUS        | NULL             | Northeast Cooling Inc. |
|         23 | CHK        | NULL             | Superior Auto Body     |
|         24 | SBL        | NULL             | AAA Insurance Inc.     |
+------------+------------+------------------+------------------------+
24 rows in set (0.00 sec)
```

In this version of the query, the individual table is outer joined to the account table within a subquery named account_ind, the results of which are then outer joined to the business table. Thus, each query (the subquery and the containing query) only uses a single outer join. If you are using a database other than MySQL, you may need to utilize this strategy if you want to outer join more than one table.

Self Outer Joins

In Chapter 5, you were introduced to the concept of the self join, where a table is joined to itself. Here's a self join example from Chapter 5, which joins the employee table to itself to generate a list of employees and their supervisors:

```
mysql> SELECT e.fname, e.lname,
    ->    e_mgr.fname mgr_fname, e_mgr.lname mgr_lname
```

```
 -> FROM employee e INNER JOIN employee e_mgr
 ->    ON e.superior_emp_id = e_mgr.emp_id;
+----------+-----------+-----------+-----------+
| fname    | lname     | mgr_fname | mgr_lname |
+----------+-----------+-----------+-----------+
| Susan    | Barker    | Michael   | Smith     |
| Robert   | Tyler     | Michael   | Smith     |
| Susan    | Hawthorne | Robert    | Tyler     |
| John     | Gooding   | Susan     | Hawthorne |
| Helen    | Fleming   | Susan     | Hawthorne |
| Chris    | Tucker    | Helen     | Fleming   |
| Sarah    | Parker    | Helen     | Fleming   |
| Jane     | Grossman  | Helen     | Fleming   |
| Paula    | Roberts   | Susan     | Hawthorne |
| Thomas   | Ziegler   | Paula     | Roberts   |
| Samantha | Jameson   | Paula     | Roberts   |
| John     | Blake     | Susan     | Hawthorne |
| Cindy    | Mason     | John      | Blake     |
| Frank    | Portman   | John      | Blake     |
| Theresa  | Markham   | Susan     | Hawthorne |
| Beth     | Fowler    | Theresa   | Markham   |
| Rick     | Tulman    | Theresa   | Markham   |
+----------+-----------+-----------+-----------+
17 rows in set (0.02 sec)
```

This query works fine except for one small issue; employees who don't have a supervisor are left out of the result set. By changing the join from an inner join to an outer join, however, the result set will include all employees, including those without supervisors:

```
mysql> SELECT e.fname, e.lname,
    ->    e_mgr.fname mgr_fname, e_mgr.lname mgr_lname
    -> FROM employee e LEFT OUTER JOIN employee e_mgr
    ->    ON e.superior_emp_id = e_mgr.emp_id;
+----------+-----------+-----------+-----------+
| fname    | lname     | mgr_fname | mgr_lname |
+----------+-----------+-----------+-----------+
| Michael  | Smith     | NULL      | NULL      |
| Susan    | Barker    | Michael   | Smith     |
| Robert   | Tyler     | Michael   | Smith     |
| Susan    | Hawthorne | Robert    | Tyler     |
| John     | Gooding   | Susan     | Hawthorne |
| Helen    | Fleming   | Susan     | Hawthorne |
| Chris    | Tucker    | Helen     | Fleming   |
| Sarah    | Parker    | Helen     | Fleming   |
| Jane     | Grossman  | Helen     | Fleming   |
| Paula    | Roberts   | Susan     | Hawthorne |
| Thomas   | Ziegler   | Paula     | Roberts   |
| Samantha | Jameson   | Paula     | Roberts   |
| John     | Blake     | Susan     | Hawthorne |
| Cindy    | Mason     | John      | Blake     |
| Frank    | Portman   | John      | Blake     |
| Theresa  | Markham   | Susan     | Hawthorne |
| Beth     | Fowler    | Theresa   | Markham   |
```

```
| Rick     | Tulman    | Theresa   | Markham   |
+----------+-----------+-----------+-----------+
18 rows in set (0.00 sec)
```

The result set now includes Michael Smith, who is the president of the bank and, therefore, does not have a supervisor. The query utilizes a left outer join to generate a list of all employees and, if applicable, their supervisor. If you change the join to be a right outer join, you would see the following results:

```
mysql> SELECT e.fname, e.lname,
    ->   e_mgr.fname mgr_fname, e_mgr.lname mgr_lname
    -> FROM employee e RIGHT OUTER JOIN employee e_mgr
    ->   ON e.superior_emp_id = e_mgr.emp_id;
+----------+-----------+-----------+-----------+
| fname    | lname     | mgr_fname | mgr_lname |
+----------+-----------+-----------+-----------+
| Susan    | Barker    | Michael   | Smith     |
| Robert   | Tyler     | Michael   | Smith     |
| NULL     | NULL      | Susan     | Barker    |
| Susan    | Hawthorne | Robert    | Tyler     |
| John     | Gooding   | Susan     | Hawthorne |
| Helen    | Fleming   | Susan     | Hawthorne |
| Paula    | Roberts   | Susan     | Hawthorne |
| John     | Blake     | Susan     | Hawthorne |
| Theresa  | Markham   | Susan     | Hawthorne |
| NULL     | NULL      | John      | Gooding   |
| Chris    | Tucker    | Helen     | Fleming   |
| Sarah    | Parker    | Helen     | Fleming   |
| Jane     | Grossman  | Helen     | Fleming   |
| NULL     | NULL      | Chris     | Tucker    |
| NULL     | NULL      | Sarah     | Parker    |
| NULL     | NULL      | Jane      | Grossman  |
| Thomas   | Ziegler   | Paula     | Roberts   |
| Samantha | Jameson   | Paula     | Roberts   |
| NULL     | NULL      | Thomas    | Ziegler   |
| NULL     | NULL      | Samantha  | Jameson   |
| Cindy    | Mason     | John      | Blake     |
| Frank    | Portman   | John      | Blake     |
| NULL     | NULL      | Cindy     | Mason     |
| NULL     | NULL      | Frank     | Portman   |
| Beth     | Fowler    | Theresa   | Markham   |
| Rick     | Tulman    | Theresa   | Markham   |
| NULL     | NULL      | Beth      | Fowler    |
| NULL     | NULL      | Rick      | Tulman    |
+----------+-----------+-----------+-----------+
28 rows in set (0.00 sec)
```

This query shows each supervisor (still the third and fourth columns) along with the set of employees he or she supervises. Therefore, Michael Smith appears twice as supervisor to Susan Barker and Robert Tyler; Susan Barker appears once as a supervisor to nobody (null values in the first and second columns). All 18 employees appear at least once in the third and fourth columns, with some appearing more than once if they supervise more than one employee, making a total of 28 rows in the result set. This is a very different outcome from the previous query, and it was prompted by changing only a single keyword (left to right). Therefore, when using outer joins (both inner and outer), make sure you think carefully about whether to specify a left or right outer join.

Cross Joins

Back in Chapter 5, I introduced the concept of a Cartesian product, which is essentially the result of joining multiple tables without specifying any join conditions. Cartesian products are used fairly frequently by accident (i.e., forgetting to add the join condition to the from clause) but are not so common otherwise. If, however, you *do* intend to generate the Cartesian product of two tables, you should specify a *cross join*, as in:

```
mysql> SELECT pt.name, p.product_cd, p.name
    -> FROM product p CROSS JOIN product_type pt;
+-----------------------------+------------+-------------------------+
| name                        | product_cd | name                    |
+-----------------------------+------------+-------------------------+
| Customer Accounts           | AUT        | auto loan               |
| Customer Accounts           | BUS        | business line of credit |
| Customer Accounts           | CD         | certificate of deposit  |
| Customer Accounts           | CHK        | checking account        |
| Customer Accounts           | MM         | money market account    |
| Customer Accounts           | MRT        | home mortgage           |
| Customer Accounts           | SAV        | savings account         |
| Customer Accounts           | SBL        | small business loan     |
| Insurance Offerings         | AUT        | auto loan               |
| Insurance Offerings         | BUS        | business line of credit |
| Insurance Offerings         | CD         | certificate of deposit  |
| Insurance Offerings         | CHK        | checking account        |
| Insurance Offerings         | MM         | money market account    |
| Insurance Offerings         | MRT        | home mortgage           |
| Insurance Offerings         | SAV        | savings account         |
| Insurance Offerings         | SBL        | small business loan     |
| Individual and Business Loans | AUT      | auto loan               |
| Individual and Business Loans | BUS      | business line of credit |
| Individual and Business Loans | CD       | certificate of deposit  |
| Individual and Business Loans | CHK      | checking account        |
| Individual and Business Loans | MM       | money market account    |
| Individual and Business Loans | MRT      | home mortgage           |
| Individual and Business Loans | SAV      | savings account         |
| Individual and Business Loans | SBL      | small business loan     |
+-----------------------------+------------+-------------------------+
24 rows in set (0.00 sec)
```

This query generates the Cartesian product of the product and product_type tables, resulting in 24 rows (8 product rows times 3 product_type rows). But now that you know what a cross join is and how to specify it, what is it used for? Most SQL books will describe what a cross join is and then tell you that it is seldom useful, but I would like to share with you a situation in which I find the cross join to be quite helpful.

Back in Chapter 9, I discussed how to use subqueries to fabricate tables. The example I used showed how to build a three-row table that could be joined to other tables. Here's the fabricated table from the example:

```
mysql> SELECT 'Small Fry' name, 0 low_limit, 4999.99 high_limit
    -> UNION ALL
    -> SELECT 'Average Joes' name, 5000 low_limit, 9999.99 high_limit
    -> UNION ALL
    -> SELECT 'Heavy Hitters' name, 10000 low_limit, 9999999.99 high_limit;
+---------------+-----------+------------+
| name          | low_limit | high_limit |
+---------------+-----------+------------+
| Small Fry     |         0 |    4999.99 |
| Average Joes  |      5000 |    9999.99 |
| Heavy Hitters |     10000 | 9999999.99 |
+---------------+-----------+------------+
3 rows in set (0.00 sec)
```

While this table was exactly what was needed for placing customers into three groups based on their aggregate account balance, this strategy of merging single-row tables using the set operator union all doesn't work very well if you need to fabricate a large table.

Say, for example, that you want to create a query that generates a row for every day in the year 2004, but you don't have a table in your database that contains a row for every day. Using the strategy from the example in Chapter 9, you could do something like the following:

```
SELECT '2004-01-01' dt
UNION ALL
SELECT '2004-01-02' dt
UNION ALL
SELECT '2004-01-03' dt
UNION ALL
...
...
...
SELECT '2004-12-29' dt
UNION ALL
SELECT '2004-12-30' dt
UNION ALL
SELECT '2004-12-31' dt
```

Building a query that merges the results of 366 queries together is a bit tedious, so maybe a different strategy is needed. What if you generate a table with 366 rows (2004 was a leap year) with a single column containing a number between 0 and 366, and then add that number of days to January 1, 2004? Here's one possible method to generate such a table:

```
mysql> SELECT ones.num + tens.num + hundreds.num
    -> FROM
```

```
    ->  (SELECT 0 num UNION ALL
    ->   SELECT 1 num UNION ALL
    ->   SELECT 2 num UNION ALL
    ->   SELECT 3 num UNION ALL
    ->   SELECT 4 num UNION ALL
    ->   SELECT 5 num UNION ALL
    ->   SELECT 6 num UNION ALL
    ->   SELECT 7 num UNION ALL
    ->   SELECT 8 num UNION ALL
    ->   SELECT 9 num) ones
    ->  CROSS JOIN
    ->  (SELECT 0 num UNION ALL
    ->   SELECT 10 num UNION ALL
    ->   SELECT 20 num UNION ALL
    ->   SELECT 30 num UNION ALL
    ->   SELECT 40 num UNION ALL
    ->   SELECT 50 num UNION ALL
    ->   SELECT 60 num UNION ALL
    ->   SELECT 70 num UNION ALL
    ->   SELECT 80 num UNION ALL
    ->   SELECT 90 num) tens
    ->  CROSS JOIN
    ->  (SELECT 0 num UNION ALL
    ->   SELECT 100 num UNION ALL
    ->   SELECT 200 num UNION ALL
    ->   SELECT 300 num) hundreds;
+----------------------------------+
| ones.num + tens.num + hundreds.num |
+----------------------------------+
|                                0 |
|                                1 |
|                                2 |
|                                3 |
|                                4 |
|                                5 |
|                                6 |
|                                7 |
|                                8 |
|                                9 |
|                               10 |
|                               11 |
|                               12 |
...
...
...
|                              391 |
|                              392 |
|                              393 |
|                              394 |
|                              395 |
|                              396 |
|                              397 |
|                              398 |
|                              399 |
```

```
+-------------------------------------+
```
400 rows in set (0.00 sec)

If you take the Cartesian product of the three sets {0, 1, 2, 3, 4, 5, 6, 7, 8, 9}, {0, 10, 20, 30, 40, 50, 60, 70, 80, 90}, and {0, 100, 200, 300} and add the values in the three columns, you get a 400 row result containing all numbers between 0 and 399. While this is more than the 366 rows needed to generate the set of days in 2004, it's easy enough to get rid of the excess rows, and I'll show you how shortly.

The next step is to convert the set of numbers to a set of dates. To do this, I will use the date_add() function to add each number in the result set to January 1, 2004. Then I'll add a filter condition to throw away any dates that venture into 2005:

```
mysql> SELECT DATE_ADD('2004-01-01',
    ->    INTERVAL (ones.num + tens.num + hundreds.num) DAY) dt
    -> FROM
    -> (SELECT 0 num UNION ALL
    ->    SELECT 1 num UNION ALL
    ->    SELECT 2 num UNION ALL
    ->    SELECT 3 num UNION ALL
    ->    SELECT 4 num UNION ALL
    ->    SELECT 5 num UNION ALL
    ->    SELECT 6 num UNION ALL
    ->    SELECT 7 num UNION ALL
    ->    SELECT 8 num UNION ALL
    ->    SELECT 9 num) ones
    -> CROSS JOIN
    -> (SELECT 0 num UNION ALL
    ->    SELECT 10 num UNION ALL
    ->    SELECT 20 num UNION ALL
    ->    SELECT 30 num UNION ALL
    ->    SELECT 40 num UNION ALL
    ->    SELECT 50 num UNION ALL
    ->    SELECT 60 num UNION ALL
    ->    SELECT 70 num UNION ALL
    ->    SELECT 80 num UNION ALL
    ->    SELECT 90 num) tens
    -> CROSS JOIN
    -> (SELECT 0 num UNION ALL
    ->    SELECT 100 num UNION ALL
    ->    SELECT 200 num UNION ALL
    ->    SELECT 300 num) hundreds
    -> WHERE DATE_ADD('2004-01-01',
    ->    INTERVAL (ones.num + tens.num + hundreds.num) DAY) < '2005-01-01';
+------------+
| dt         |
+------------+
| 2004-01-01 |
| 2004-01-02 |
| 2004-01-03 |
| 2004-01-04 |
| 2004-01-05 |
| 2004-01-06 |
```

```
| 2004-01-07 |
| 2004-01-08 |
| 2004-01-09 |
| 2004-01-10 |
...
...
...
| 2004-02-20 |
| 2004-02-21 |
| 2004-02-22 |
| 2004-02-23 |
| 2004-02-24 |
| 2004-02-25 |
| 2004-02-26 |
| 2004-02-27 |
| 2004-02-28 |
| 2004-02-29 |
| 2004-03-01 |
...
...
...
| 2004-12-20 |
| 2004-12-21 |
| 2004-12-22 |
| 2004-12-23 |
| 2004-12-24 |
| 2004-12-25 |
| 2004-12-26 |
| 2004-12-27 |
| 2004-12-28 |
| 2004-12-29 |
| 2004-12-30 |
| 2004-12-31 |
+------------+
366 rows in set (0.01 sec)
```

The nice thing about this approach is that the result set automatically includes the extra leap day (February 29) without your intervention, since the database server figures it out when it adds 59 days to January 1, 2004.

Now that you have a mechanism for fabricating all of the days in 2004, what should you do with it? Well, you might be asked to generate a query that shows every day in 2004 along with the number of accounts opened on that day, the number of transactions posted on that day, etc. Here's an example that answers the first question:

```
mysql> SELECT days.dt, COUNT(a.account_id)
    -> FROM account a RIGHT OUTER JOIN
    ->   (SELECT DATE_ADD('2004-01-01',
    ->      INTERVAL (ones.num + tens.num + hundreds.num) DAY) dt
    ->   FROM
    ->     (SELECT 0 num UNION ALL
    ->       SELECT 1 num UNION ALL
    ->       SELECT 2 num UNION ALL
```

```
    ->      SELECT 3 num UNION ALL
    ->      SELECT 4 num UNION ALL
    ->      SELECT 5 num UNION ALL
    ->      SELECT 6 num UNION ALL
    ->      SELECT 7 num UNION ALL
    ->      SELECT 8 num UNION ALL
    ->      SELECT 9 num) ones
    ->    CROSS JOIN
    ->    (SELECT 0 num UNION ALL
    ->      SELECT 10 num UNION ALL
    ->      SELECT 20 num UNION ALL
    ->      SELECT 30 num UNION ALL
    ->      SELECT 40 num UNION ALL
    ->      SELECT 50 num UNION ALL
    ->      SELECT 60 num UNION ALL
    ->      SELECT 70 num UNION ALL
    ->      SELECT 80 num UNION ALL
    ->      SELECT 90 num) tens
    ->    CROSS JOIN
    ->    (SELECT 0 num UNION ALL
    ->      SELECT 100 num UNION ALL
    ->      SELECT 200 num UNION ALL
    ->      SELECT 300 num) hundreds
    ->    WHERE DATE_ADD('2004-01-01',
    ->      INTERVAL (ones.num + tens.num + hundreds.num) DAY) <
    ->        '2005-01-01') days
    ->    ON days.dt = a.open_date
    -> GROUP BY days.dt;
+------------+---------------------+
| dt         | COUNT(a.account_id) |
+------------+---------------------+
| 2004-01-01 |                   0 |
| 2004-01-02 |                   0 |
| 2004-01-03 |                   0 |
| 2004-01-04 |                   0 |
| 2004-01-05 |                   0 |
| 2004-01-06 |                   0 |
| 2004-01-07 |                   0 |
| 2004-01-08 |                   0 |
| 2004-01-09 |                   0 |
| 2004-01-10 |                   0 |
| 2004-01-11 |                   0 |
| 2004-01-12 |                   1 |
| 2004-01-13 |                   0 |
| 2004-01-14 |                   0 |
| 2004-01-15 |                   0 |
...
...
...
| 2004-12-15 |                   0 |
| 2004-12-16 |                   0 |
| 2004-12-17 |                   0 |
| 2004-12-18 |                   0 |
| 2004-12-19 |                   0 |
```

```
|  2004-12-20 |                   0 |
|  2004-12-21 |                   0 |
|  2004-12-22 |                   0 |
|  2004-12-23 |                   0 |
|  2004-12-24 |                   0 |
|  2004-12-25 |                   0 |
|  2004-12-26 |                   0 |
|  2004-12-27 |                   0 |
|  2004-12-28 |                   1 |
|  2004-12-29 |                   0 |
|  2004-12-30 |                   0 |
|  2004-12-31 |                   0 |
+-------------+---------------------+
366 rows in set (0.03 sec)
```

This is one of the more interesting queries thus far in the book, in that it includes cross joins, outer joins, a date function, grouping, set operations (union all), and an aggregate function (count( )). It is also not the most elegant solution to the given problem, but it should serve as an example of how, with a little creativity and a firm grasp of the language, you can make even a seldom-used feature like cross joins a potent tool in your SQL toolkit.

Natural Joins

If you are lazy (and aren't we all), you can choose a join type that allows you to name the tables to be joined but lets the database server determine what the join conditions need to be. Known as the *natural join*, this join type relies on identical column names across multiple tables to infer the proper join conditions. For example, the account table includes a column named cust_id, which is the foreign key to the customer table, whose primary key is also named cust_id. Thus, you can write a query that uses a natural join to join the two tables:

```
mysql> SELECT a.account_id, a.cust_id, c.cust_type_cd, c.fed_id
    -> FROM account a NATURAL JOIN customer c;
+------------+---------+--------------+-------------+
| account_id | cust_id | cust_type_cd | fed_id      |
+------------+---------+--------------+-------------+
|          1 |       1 | I            | 111-11-1111 |
|          2 |       1 | I            | 111-11-1111 |
|          3 |       1 | I            | 111-11-1111 |
|          4 |       2 | I            | 222-22-2222 |
|          5 |       2 | I            | 222-22-2222 |
|          6 |       3 | I            | 333-33-3333 |
|          7 |       3 | I            | 333-33-3333 |
|          8 |       4 | I            | 444-44-4444 |
|          9 |       4 | I            | 444-44-4444 |
|         10 |       4 | I            | 444-44-4444 |
|         11 |       5 | I            | 555-55-5555 |
|         12 |       6 | I            | 666-66-6666 |
|         13 |       6 | I            | 666-66-6666 |
```

```
|          14 |        7 | I |      | 777-77-7777 |
|          15 |        8 | I |      | 888-88-8888 |
|          16 |        8 | I |      | 888-88-8888 |
|          17 |        9 | I |      | 999-99-9999 |
|          18 |        9 | I |      | 999-99-9999 |
|          19 |        9 | T |      | 999-99-9999 |
|          20 |       10 | B |      | 04-1111111  |
|          21 |       10 | B |      | 04-1111111  |
|          22 |       11 | B |      | 04-2222222  |
|          23 |       12 | B |      | 04-3333333  |
|          24 |       13 | B |      | 04-4444444  |
+-------------+----------+---------------+-------------+
24 rows in set (0.02 sec)
```

Because you specified a natural join, the server inspected the table definitions and added the join condition a.cust_id = c.cust_id to join the two tables.

This is all well and good, but what if the columns don't have the same name across the tables? For example, the account table also has a foreign key to the branch table, but the column in the account table is named open_branch_id instead of just branch_id. Let's see what happens if I try a natural join between the account and branch tables:

```
mysql> SELECT a.account_id, a.cust_id, a.open_branch_id,
    -> FROM account a NATURAL JOIN branch b;
+------------+---------+----------------+--------------+
| account_id | cust_id | open_branch_id | name         |
+------------+---------+----------------+--------------+
|          1 |       1 |              2 | Headquarters |
|          2 |       1 |              2 | Headquarters |
|          3 |       1 |              2 | Headquarters |
|          4 |       2 |              2 | Headquarters |
|          5 |       2 |              2 | Headquarters |
|          6 |       3 |              3 | Headquarters |
|          7 |       3 |              3 | Headquarters |
|          8 |       4 |              1 | Headquarters |
|          9 |       4 |              1 | Headquarters |
|         10 |       4 |              1 | Headquarters |
|         11 |       5 |              4 | Headquarters |
|         12 |       6 |              1 | Headquarters |
|         13 |       6 |              1 | Headquarters |
|         14 |       7 |              2 | Headquarters |
|         15 |       8 |              4 | Headquarters |
|         16 |       8 |              4 | Headquarters |
|         17 |       9 |              1 | Headquarters |
|         18 |       9 |              1 | Headquarters |
|         19 |       9 |              1 | Headquarters |
|         20 |      10 |              4 | Headquarters |
|         21 |      10 |              4 | Headquarters |
|         22 |      11 |              2 | Headquarters |
|         23 |      12 |              4 | Headquarters |
|         24 |      13 |              3 | Headquarters |
...
...
...
```

```
|           1 |       1 |             2 | So. NH Branch |
|           2 |       1 |             2 | So. NH Branch |
|           3 |       1 |             2 | So. NH Branch |
|           4 |       2 |             2 | So. NH Branch |
|           5 |       2 |             2 | So. NH Branch |
|           6 |       3 |             3 | So. NH Branch |
|           7 |       3 |             3 | So. NH Branch |
|           8 |       4 |             1 | So. NH Branch |
|           9 |       4 |             1 | So. NH Branch |
|          10 |       4 |             1 | So. NH Branch |
|          11 |       5 |             4 | So. NH Branch |
|          12 |       6 |             1 | So. NH Branch |
|          13 |       6 |             1 | So. NH Branch |
|          14 |       7 |             2 | So. NH Branch |
|          15 |       8 |             4 | So. NH Branch |
|          16 |       8 |             4 | So. NH Branch |
|          17 |       9 |             1 | So. NH Branch |
|          18 |       9 |             1 | So. NH Branch |
|          19 |       9 |             1 | So. NH Branch |
|          20 |      10 |             4 | So. NH Branch |
|          21 |      10 |             4 | So. NH Branch |
|          22 |      11 |             2 | So. NH Branch |
|          23 |      12 |             4 | So. NH Branch |
|          24 |      13 |             3 | So. NH Branch |
+-------------+---------+---------------+---------------+
96 rows in set (0.03 sec)
```

It looks like something has gone wrong; the query should return no more than 24
rows, since there are 24 rows in the account table. What has happened is that, since
the server couldn't find two identically named columns in the two tables, no join
condition was generated and the two tables were cross-joined instead, resulting in 96
rows (24 accounts times 4 branches).

So, is the reduced wear and tear on the old fingers from not having to type the join
condition worth the trouble? Absolutely not; you should avoid this join type and use
inner joins with explicit join conditions.

Exercises

The following exercises will test your understanding of outer and cross joins. Please
see Appendix C for solutions.

10-1

Write a query that returns all product names along with the accounts based on that
product (use the product_cd column in the account table to link to the product table).
Include all products, even if no accounts have been opened for that product.

10-2

Reformulate your query from exercise 10-1 to use the other outer join type (i.e., if you used a left outer join in 10-1, use a right outer join this time) such that the results are identical to 10-1.

10-3

Outer join the account table to both the individual and business tables (via the account.cust_id column) such that the result set contains one row per account. Columns to include are account.account_id, account.product_cd, individual.fname, individual.lname, and business.name.

10-4 (Extra Credit)

Devise a query that will generate the set {1, 2, 3,…, 99, 100}. (Hint: Use a cross join with at least two from clause subqueries.)

CHAPTER 11

Conditional Logic

In certain situations, you may want your SQL statements to take one course of action or another depending on the values of certain columns or expressions. This chapter focuses on how to write statements that can behave differently depending on the data encountered during statement execution.

What Is Conditional Logic?

Conditional logic is simply the ability to take one of several paths during program execution. For example, when querying customer information, you might want to retrieve either the fname/lname columns from the individual table or the name column from the business table depending on what type of customer is encountered. Using outer joins, you could return both strings and let the caller figure out which one to use, as in:

```
mysql> SELECT c.cust_id, c.fed_id, c.cust_type_cd,
    ->    CONCAT(i.fname, ' ', i.lname) indiv_name,
    ->    b.name business_name
    -> FROM customer c LEFT OUTER JOIN individual i
    ->    ON c.cust_id = i.cust_id
    ->    LEFT OUTER JOIN business b
    ->    ON c.cust_id = b.cust_id;
+---------+-------------+--------------+----------------+---------------------+
| cust_id | fed_id      | cust_type_cd | indiv_name     | business_name       |
+---------+-------------+--------------+----------------+---------------------+
|       1 | 111-11-1111 | I            | James Hadley   | NULL                |
|       2 | 222-22-2222 | I            | Susan Tingley  | NULL                |
|       3 | 333-33-3333 | I            | Frank Tucker   | NULL                |
|       4 | 444-44-4444 | I            | John Hayward   | NULL                |
|       5 | 555-55-5555 | I            | Charles Frasier| NULL                |
|       6 | 666-66-6666 | I            | John Spencer   | NULL                |
|       7 | 777-77-7777 | I            | Margaret Young | NULL                |
|       8 | 888-88-8888 | I            | Louis Blake    | NULL                |
|       9 | 999-99-9999 | I            | Richard Farley | NULL                |
|      10 | 04-1111111  | B            | NULL           | Chilton Engineering |
```

```
|       11 | 04-2222222  | B            | NULL            | Northeast Cooling Inc. |
|       12 | 04-3333333  | B            | NULL            | Superior Auto Body     |
|       13 | 04-4444444  | B            | NULL            | AAA Insurance Inc.     |
+----------+-------------+--------------+-----------------+------------------------+
13 rows in set (0.13 sec)
```

The caller can look at the value of the cust_type_cd column and decide whether to use the indiv_name or business_name column. Instead, however, you could use conditional logic via a *case expression* to determine the type of customer and return the appropriate string, as in:

```
mysql> SELECT c.cust_id, c.fed_id,
    ->    CASE
    ->      WHEN c.cust_type_cd = 'I'
    ->        THEN CONCAT(i.fname, ' ', i.lname)
    ->      WHEN c.cust_type_cd = 'B'
    ->        THEN b.name
    ->      ELSE 'Unknown'
    ->    END name
    -> FROM customer c LEFT OUTER JOIN individual i
    ->    ON c.cust_id = i.cust_id
    ->    LEFT OUTER JOIN business b
    ->    ON c.cust_id = b.cust_id;
+---------+-------------+------------------------+
| cust_id | fed_id      | name                   |
+---------+-------------+------------------------+
|       1 | 111-11-1111 | James Hadley           |
|       2 | 222-22-2222 | Susan Tingley          |
|       3 | 333-33-3333 | Frank Tucker           |
|       4 | 444-44-4444 | John Hayward           |
|       5 | 555-55-5555 | Charles Frasier        |
|       6 | 666-66-6666 | John Spencer           |
|       7 | 777-77-7777 | Margaret Young         |
|       8 | 888-88-8888 | Louis Blake            |
|       9 | 999-99-9999 | Richard Farley         |
|      10 | 04-1111111  | Chilton Engineering    |
|      11 | 04-2222222  | Northeast Cooling Inc. |
|      12 | 04-3333333  | Superior Auto Body     |
|      13 | 04-4444444  | AAA Insurance Inc.     |
+---------+-------------+------------------------+
13 rows in set (0.00 sec)
```

This version of the query returns a single name column that is generated by the *case expression starting on the second line of the query*, which, in this case, checks the value of the cust_type_cd column and returns either the individual's first/last names *or* the business name.

The Case Expression

All of the major database servers include built-in functions designed to mimic the if-then-else statement found in most programming languages (examples include Oracle's decode( ) function, MySQL's if( ) function, and SQL Server's coalesce( ) function). Case expressions are also designed to facilitate if-then-else logic but enjoy two advantages over built-in functions:

- The case expression is part of the SQL standard (SQL92 release) and has been implemented by Oracle Database, SQL Server, and MySQL.
- Case expressions are built into the SQL grammar and can be included in select, insert, update, and delete statements.

The next two subsections will introduce the two different types of case expressions, and then I will show you some examples of case expressions in action.

Searched Case Expressions

The case expression demonstrated earlier in the chapter is an example of a *searched case expression*, which has the following syntax:

```
CASE
  WHEN C1 THEN E1
  WHEN C2 THEN E2
  ...
  WHEN CN THEN EN
  [ELSE ED]
END
```

In the previous definition, the symbols C1, C2,..., CN represent conditions, and the symbols E1, E2,..., EN represent expressions to be returned by the case expression. If the condition in a when clause evaluates to true, then the corresponding expression is returned by the case expression. Additionally, the ED symbol represents the default expression, which is returned by the case expression if *none* of the conditions C1, C2,..., CN evaluate to true (the else clause is optional, which is why it is enclosed in square brackets). All of the expressions returned by the various when clauses must evaluate to the same type (e.g., date, number, varchar).

Here's an example of a searched case expression:

```
CASE
  WHEN employee.title = 'Head Teller'
    THEN 'Head Teller'
  WHEN employee.title = 'Teller'
    AND YEAR(employee.start_date) > 2004
    THEN 'Teller Trainee'
  WHEN employee.title = 'Teller'
    AND YEAR(employee.start_date) < 2003
    THEN 'Experienced Teller'
  WHEN employee.title = 'Teller'
    THEN 'Teller'
  ELSE 'Non-Teller'
END
```

This case expression returns a string that can be used to determine hourly pay scales, print name badges, etc. When the case expression is evaluated, the when clauses are evaluated in order from top to bottom; as soon as one of the conditions in a when clause evaluates to true, the corresponding expression is returned and any remaining when clauses are ignored. If none of the when clause conditions evaluate to true, then the expression in the else clause is returned.

Although the previous example returns string expressions, keep in mind that any type of expression may be returned by case expressions, including subqueries. Here's another version of the individual/business name query from earlier in the chapter that uses subqueries instead of outer joins to retrieve data from the individual and business tables:

```
mysql> SELECT c.cust_id, c.fed_id,
    ->    CASE
    ->      WHEN c.cust_type_cd = 'I' THEN
    ->        (SELECT CONCAT(i.fname, ' ', i.lname)
    ->         FROM individual i
    ->         WHERE i.cust_id = c.cust_id)
    ->      WHEN c.cust_type_cd = 'B' THEN
    ->        (SELECT b.name
    ->         FROM business b
    ->         WHERE b.cust_id = c.cust_id)
    ->      ELSE 'Unknown'
    ->    END name
    -> FROM customer c;
+---------+-------------+------------------------+
| cust_id | fed_id      | name                   |
+---------+-------------+------------------------+
|       1 | 111-11-1111 | James Hadley           |
|       2 | 222-22-2222 | Susan Tingley          |
|       3 | 333-33-3333 | Frank Tucker           |
|       4 | 444-44-4444 | John Hayward           |
|       5 | 555-55-5555 | Charles Frasier        |
|       6 | 666-66-6666 | John Spencer           |
|       7 | 777-77-7777 | Margaret Young         |
|       8 | 888-88-8888 | Louis Blake            |
|       9 | 999-99-9999 | Richard Farley         |
|      10 | 04-1111111  | Chilton Engineering    |
|      11 | 04-2222222  | Northeast Cooling Inc. |
|      12 | 04-3333333  | Superior Auto Body     |
|      13 | 04-4444444  | AAA Insurance Inc.     |
+---------+-------------+------------------------+
13 rows in set (0.01 sec)
```

This version of the query includes only the customer table in the from clause and uses correlated subqueries to retrieve the appropriate name for each customer. I prefer this version over the outer-join version from earlier in the chapter, since the server only reads from the individual and business tables as needed instead of joining all three tables.

Simple Case Expressions

The *simple case expression* is quite similar to the searched case expression but is a bit less functional. Here's the syntax:

```
CASE V0
  WHEN V1 THEN E1
  WHEN V2 THEN E2
  ...
  WHEN VN THEN EN
  [ELSE ED]
END
```

In the above definition, V0 represents a value, and the symbols V1, V2,..., VN represent values that are to be compared to V0. The symbols E1, E2,..., EN represent expressions to be returned by the case expression, and ED represents the expression to be returned if none of the values in the set V1, V2,..., VN match the V0 value.

Here's an example of a simple case expression:

```
CASE customer.cust_type_cd
  WHEN 'I' THEN
    (SELECT CONCAT(i.fname, ' ', i.lname)
     FROM individual I
     WHERE i.cust_id = customer.cust_id)
  WHEN 'B' THEN
    (SELECT b.name
     FROM business b
     WHERE b.cust_id = customer.cust_id)
  ELSE 'Unknown Customer Type'
END
```

Simple case expressions are less functional than searched case expressions because you can't specify your own conditions; instead, equality conditions are built for you. To show you what I mean, here's a searched case expression having the same logic as the previous simple case expression:

```
CASE
  WHEN customer.cust_type_cd = 'I' THEN
    (SELECT CONCAT(i.fname, ' ', i.lname)
     FROM individual I
     WHERE i.cust_id = customer.cust_id)
  WHEN customer.cust_type_cd = 'B' THEN
    (SELECT b.name
     FROM business b
     WHERE b.cust_id = customer.cust_id)
  ELSE 'Unknown Customer Type'
END
```

With searched case expressions, you can build range conditions, inequality conditions, and multipart conditions using and/or/not, so I would recommend using searched case expressions for all but the simplest logic.

Case Expression Examples

The following sections present a variety of examples illustrating the utility of conditional logic in SQL statements.

Result Set Transformations

You may have run into a situation where you are performing aggregations over a finite set of values, such as days of the week, but you want the result set to contain a single row with one column per value instead of one row per value. As an example, let's say you have a query that returns the number of accounts opened in each year since 2000:

```
mysql> SELECT YEAR(open_date) year, COUNT(*) how_many
    -> FROM account
    -> WHERE open_date > '1999-12-31'
    -> GROUP BY YEAR(open_date);
+------+----------+
| year | how_many |
+------+----------+
| 2000 |        3 |
| 2001 |        4 |
| 2002 |        5 |
| 2003 |        3 |
| 2004 |        9 |
+------+----------+
5 rows in set (0.00 sec)
```

To transform this result set into a single row with six columns (one column for each year from 2000 to 2005), you will need to create six columns and, within each column, sum *only* those rows pertaining to the year in question:

```
mysql> SELECT
    ->   SUM(CASE
    ->       WHEN EXTRACT(YEAR FROM open_date) = 2000 THEN 1
    ->       ELSE 0
    ->     END) year_2000,
    ->   SUM(CASE
    ->       WHEN EXTRACT(YEAR FROM open_date) = 2001 THEN 1
    ->       ELSE 0
    ->     END) year_2001,
    ->   SUM(CASE
    ->       WHEN EXTRACT(YEAR FROM open_date) = 2002 THEN 1
    ->       ELSE 0
    ->     END) year_2002,
    ->   SUM(CASE
    ->       WHEN EXTRACT(YEAR FROM open_date) = 2003 THEN 1
    ->       ELSE 0
    ->     END) year_2003,
    ->   SUM(CASE
    ->       WHEN EXTRACT(YEAR FROM open_date) = 2004 THEN 1
    ->       ELSE 0
```

```
   ->        END) year_2004,
   ->    SUM(CASE
   ->          WHEN EXTRACT(YEAR FROM open_date) = 2005 THEN 1
   ->          ELSE 0
   ->          END) year_2005
   -> FROM account
   -> WHERE open_date > '1999-12-31';
+-----------+-----------+-----------+-----------+-----------+-----------+
| year_2000 | year_2001 | year_2002 | year_2003 | year_2004 | year_2005 |
+-----------+-----------+-----------+-----------+-----------+-----------+
|         3 |         4 |         5 |         3 |         9 |         0 |
+-----------+-----------+-----------+-----------+-----------+-----------+
1 row in set (0.01 sec)
```

Each of the six columns in the previous query is identical, except for the year value. When the extract( ) function returns the desired year, then the value 1 is returned by the case expression; otherwise, a 0 is returned. When summed over all accounts opened since 2000, each column returns the number of accounts opened for that year. Obviously, such transformations are only practical for a small number of values; generating one column for each year since 1905 would quickly become tedious.

Selective Aggregation

Back in Chapter 9, I showed a partial solution for an example that demonstrated how to find accounts whose account balances don't agree with the raw data in the transaction table. The reason for the partial solution was that a full solution requires the use of conditional logic, so all the pieces are now in place to finish the job. Here's where I left off from Chapter 9:

```
SELECT CONCAT('ALERT! : Account #', a.account_id,
  ' Has Incorrect Balance!')
FROM account a
WHERE (a.avail_balance, a.pending_balance) <>
 (SELECT SUM(<expression to generate available balance>),
    SUM(<expression to generate pending balance>)
  FROM transaction t
  WHERE t.account_id = a.account_id);
```

The query uses a correlated subquery on the transaction table to sum together the individual transactions for a given account. When summing transactions, there are two issues that need to be considered:

- Transaction amounts are always positive, so you need to look at the transaction type to see whether the transaction is a debit or a credit and flip the sign (multiply by −1) for debit transactions.

- If the date in the funds_avail_date column is greater than the current day, then the transaction should be added to the pending balance total but not to the available balance total.

While some transactions need to be excluded from the available balance, all transactions are included in the pending balance, making it the simpler of the two calculations. Here's the case expression used to calculate pending balance:

```
CASE
  WHEN transaction.txn_type_cd = 'DBT'
    THEN transaction.amount * -1
  ELSE transaction.amount
END
```

Thus, all transaction amounts are multiplied by −1 for debit transactions and left as is for credit transactions. This same logic applies to the available balance calculation as well, but only transactions that have become available should be included. Therefore, the case expression used to calculate available balance includes one additional when clause:

```
CASE
  WHEN transaction.funds_avail_date > CURRENT_TIMESTAMP()
    THEN 0
  WHEN transaction.txn_type_cd = 'DBT'
    THEN transaction.amount * -1
  ELSE transaction.amount
END
```

With the first when clause in place, unavailable funds, such as checks that have not cleared, will contribute $0 to the sum. Here's the final query with the two case expressions in place:

```
SELECT CONCAT('ALERT! : Account #', a.account_id,
  ' Has Incorrect Balance!')
FROM account a
WHERE (a.avail_balance, a.pending_balance) <>
 (SELECT
    SUM(CASE
          WHEN t.funds_avail_date > CURRENT_TIMESTAMP()
            THEN 0
          WHEN t.txn_type_cd = 'DBT'
            THEN t.amount * -1
          ELSE t.amount
        END),
    SUM(CASE
          WHEN t.txn_type_cd = 'DBT'
            THEN t.amount * -1
          ELSE t.amount
        END)
  FROM transaction t
  WHERE t.account_id = a.account_id);
```

By using conditional logic, the sum() aggregate functions are being fed manipulated data by the two case expressions, allowing the appropriate amounts to be summed.

Checking for Existence

In some cases, you want to determine whether a relationship exists between two entities without regard for the quantity. For example, you might want to know whether a customer has any checking or savings accounts, but you don't care whether a customer has more than one of each type of account. Here's a query that uses multiple case expressions to generate two output columns, one to show whether the customer has any checking accounts, and the other to show whether the customer has any savings accounts:

```
mysql> SELECT c.cust_id, c.fed_id, c.cust_type_cd,
    ->    CASE
    ->      WHEN EXISTS (SELECT 1 FROM account a
    ->        WHERE a.cust_id = c.cust_id
    ->          AND a.product_cd = 'CHK') THEN 'Y'
    ->      ELSE 'N'
    ->    END has_checking,
    ->    CASE
    ->      WHEN EXISTS (SELECT 1 FROM account a
    ->        WHERE a.cust_id = c.cust_id
    ->          AND a.product_cd = 'SAV') THEN 'Y'
    ->      ELSE 'N'
    ->    END has_savings
    -> FROM customer c;
+---------+-------------+--------------+--------------+-------------+
| cust_id | fed_id      | cust_type_cd | has_checking | has_savings |
+---------+-------------+--------------+--------------+-------------+
|       1 | 111-11-1111 | I            | Y            | Y           |
|       2 | 222-22-2222 | I            | Y            | Y           |
|       3 | 333-33-3333 | I            | Y            | N           |
|       4 | 444-44-4444 | I            | Y            | Y           |
|       5 | 555-55-5555 | I            | Y            | N           |
|       6 | 666-66-6666 | I            | Y            | N           |
|       7 | 777-77-7777 | I            | N            | N           |
|       8 | 888-88-8888 | I            | Y            | Y           |
|       9 | 999-99-9999 | I            | Y            | N           |
|      10 | 04-1111111  | B            | Y            | N           |
|      11 | 04-2222222  | B            | N            | N           |
|      12 | 04-3333333  | B            | Y            | N           |
|      13 | 04-4444444  | B            | N            | N           |
+---------+-------------+--------------+--------------+-------------+
13 rows in set (0.00 sec)
```

Each case expression includes a correlated subquery against the account table; one looks for checking accounts, the other for savings accounts. Since each when clause uses the exists operator, the conditions evaluate to true as long as the customer has at least one of the desired accounts.

In other cases, you may care about how many rows are encountered, but only up to a point. For example, the next query uses a simple case expression to count the number of accounts for each customer, and then returns either 'None', '1', '2', or '3+':

```
mysql> SELECT c.cust_id, c.fed_id, c.cust_type_cd,
    ->   CASE (SELECT COUNT(*) FROM account a
    ->     WHERE a.cust_id = c.cust_id)
    ->     WHEN 0 THEN 'None'
    ->     WHEN 1 THEN '1'
    ->     WHEN 2 THEN '2'
    ->     ELSE '3+'
    ->   END num_accounts
    -> FROM customer c;
+---------+-------------+--------------+--------------+
| cust_id | fed_id      | cust_type_cd | num_accounts |
+---------+-------------+--------------+--------------+
|       1 | 111-11-1111 | I            | 3+           |
|       2 | 222-22-2222 | I            | 2            |
|       3 | 333-33-3333 | I            | 2            |
|       4 | 444-44-4444 | I            | 3+           |
|       5 | 555-55-5555 | I            | 1            |
|       6 | 666-66-6666 | I            | 2            |
|       7 | 777-77-7777 | I            | 1            |
|       8 | 888-88-8888 | I            | 2            |
|       9 | 999-99-9999 | I            | 3+           |
|      10 | 04-1111111  | B            | 2            |
|      11 | 04-2222222  | B            | 1            |
|      12 | 04-3333333  | B            | 1            |
|      13 | 04-4444444  | B            | 1            |
+---------+-------------+--------------+--------------+
13 rows in set (0.01 sec)
```

For this query, I didn't want to differentiate between customers having more than two accounts, so the case expression simply creates a '3+' category. Such a query might be useful if you were looking for customers to contact regarding opening a new account with the bank.

Division by Zero Errors

When performing calculations that include division, you should always take care to ensure that the denominators are never equal to zero. Whereas some database servers, such as Oracle Database, will throw an error when a zero denominator is encountered, MySQL simply sets the result of the calculation to null, as demonstrated by the following:

```
mysql> SELECT 100 / 0;
+---------+
| 100 / 0 |
+---------+
|    NULL |
+---------+
1 row in set (0.00 sec)
```

To safeguard your calculations from encountering errors or, even worse, from being mysteriously set to null, you should wrap all denominators in conditional logic, as demonstrated by the following:

```
mysql> SELECT a.cust_id, a.product_cd, a.avail_balance /
    ->   CASE
    ->     WHEN prod_tots.tot_balance = 0 THEN 1
    ->     ELSE prod_tots.tot_balance
    ->   END percent_of_total
    -> FROM account a INNER JOIN
    ->   (SELECT a.product_cd, SUM(a.avail_balance) tot_balance
    ->    FROM account a
    ->    GROUP BY a.product_cd) prod_tots
    ->   ON a.product_cd = prod_tots.product_cd;
+---------+------------+------------------+
| cust_id | product_cd | percent_of_total |
+---------+------------+------------------+
|      10 | BUS        |           0.0000 |
|      11 | BUS        |           1.0000 |
|       1 | CD         |           0.1538 |
|       6 | CD         |           0.5128 |
|       7 | CD         |           0.2564 |
|       9 | CD         |           0.0769 |
|       1 | CHK        |           0.0145 |
|       2 | CHK        |           0.0309 |
|       3 | CHK        |           0.0145 |
|       4 | CHK        |           0.0073 |
|       5 | CHK        |           0.0307 |
|       6 | CHK        |           0.0017 |
|       8 | CHK        |           0.0478 |
|       9 | CHK        |           0.0017 |
|      10 | CHK        |           0.3229 |
|      12 | CHK        |           0.5281 |
|       3 | MM         |           0.1298 |
|       4 | MM         |           0.3219 |
|       9 | MM         |           0.5483 |
|       1 | SAV        |           0.2694 |
|       2 | SAV        |           0.1078 |
|       4 | SAV        |           0.4137 |
|       8 | SAV        |           0.2091 |
|      13 | SBL        |           1.0000 |
+---------+------------+------------------+
24 rows in set (0.00 sec)
```

This query computes the ratio of each account balance to the total balance for all accounts of the same product type. Since some product types, such as business loans, could have a total balance of zero if all loans were currently paid in full, it is best to include the case expression to ensure that the denominator is never zero.

Conditional Updates

When updating rows in a table, you sometimes need to make decisions regarding what values to set certain columns to. For example, after inserting a new transaction, you need to modify the avail_balance, pending_balance, and last_activity_date columns in the account table. While the last two columns are easily updated, to correctly modify the avail_balance column, you need to know whether the funds from the transaction are immediately available by checking the funds_avail_date column in the transaction table. Given that transaction ID 999 has just been inserted, the following update statement can be used to modify the three columns in the account table:

```
1   UPDATE account
2     SET last_activity_date = CURRENT_TIMESTAMP(),
3     pending_balance = pending_balance +
4       (SELECT t.amount *
5         CASE t.txn_type_cd WHEN 'DBT' THEN -1 ELSE 1 END
6       FROM transaction t
7       WHERE t.txn_id = 999),
8     avail_balance = avail_balance +
9       (SELECT
10          CASE
11            WHEN t.funds_avail_date > CURRENT_TIMESTAMP() THEN 0
12            ELSE t.amount *
13              CASE t.txn_type_cd WHEN 'DBT' THEN -1 ELSE 1 END
14          END
15        FROM transaction t
16        WHERE t.txn_id = 999)
17   WHERE account.account_id =
18     (SELECT t.account_id
19      FROM transaction t
20      WHERE t.txn_id = 999);
```

There are a total of three case expressions in this statement: two of them (lines 5 and 13) are used to flip the sign on the transaction amount for debit transactions, and the third case expression (line 10) is used to check the funds availability date. If the date is in the future, then zero is added to the available balance; otherwise, the transaction amount is added.

Handling Null Values

While null values are the appropriate thing to store in a table if the value for a column is unknown, it is not always appropriate to retrieve null values for display or to take part in expressions. For example, you might want to display the word "unknown" on a data entry screen rather than leaving a field blank. When retrieving the data, you can use a case expression to substitute the string if the value is null, as in:

```
SELECT emp_id, fname, lname,
  CASE
    WHEN title IS NULL THEN 'Unknown'
```

```
          ELSE title
      END
    FROM employee;
```

For calculations, null values often cause a null result, as demonstrated by the following:

```
mysql> SELECT (7 * 5) / ((3 + 14) * null);
+-----------------------------+
| (7 * 5) / ((3 + 14) * null) |
+-----------------------------+
|                        NULL |
+-----------------------------+
1 row in set (0.08 sec)
```

When performing calculations, case expressions are useful for translating a null value into a number (usually 0 or 1) that will allow the calculation to yield a non-null value. If you are performing a calculation that includes the account.avail_balance column, for example, you could substitute a 0 (if doing addition or subtraction) or a 1 (if doing multiplication or division) for those accounts that have been established but haven't yet been funded:

```
SELECT <some calcuation> +
  CASE
    WHEN avail_balance IS NULL THEN 0
    ELSE avail_balance
  END
  + <rest of calculation>
...
```

If a numeric column is allowed to contain null values, it is generally a good idea to use conditional logic in any calculations that include the column so that the results are usable.

Exercises

Challenge your ability to work through conditional logic problems with the examples that follow. When you're done, compare your solutions with those in Appendix C.

11-1

Rewrite the following query, which uses a simple case expression, so that the same results are achieved using a searched case expression. Try to use as few when clauses as possible.

```
SELECT emp_id,
  CASE title
    WHEN 'President' THEN 'Management'
    WHEN 'Vice President' THEN 'Management'
    WHEN 'Treasurer' THEN 'Management'
```

```
      WHEN 'Loan Manager' THEN 'Management'
      WHEN 'Operations Manager' THEN 'Operations'
      WHEN 'Head Teller' THEN 'Operations'
      WHEN 'Teller' THEN 'Operations'
      ELSE 'Unknown'
  END
FROM employee;
```

11-2

Rewrite the following query so that the result set contains a single row with four columns (one for each branch). Name the four columns branch_1 through branch_4.

```
mysql> SELECT open_branch_id, COUNT(*)
    -> FROM account
    -> GROUP BY open_branch_id;
+----------------+----------+
| open_branch_id | COUNT(*) |
+----------------+----------+
|              1 |        8 |
|              2 |        7 |
|              3 |        3 |
|              4 |        6 |
+----------------+----------+
4 rows in set (0.00 sec)
```

CHAPTER 12

Transactions

All of the examples thus far in this book have been individual SQL statements. This chapter explores the need and the infrastructure necessary to execute multiple SQL statements together.

Multiuser Databases

Database management systems allow not only a single user to query and modify data, but multiple people to do so simultaneously. If every user is executing queries, such as might be the case with a data warehouse during normal business hours, then there are very few issues for the database server to deal with. If some of the users are adding and/or modifying data, however, then there is quite a bit more bookkeeping to be done by the server.

Let's say, for example, that you are running a report that shows the available balance for all of the checking accounts opened at your branch. At the same time you are running the report, however, the following activities are occurring:

- A teller at your branch is handling a deposit for one of your customers.
- A customer is finishing a withdrawal at the ATM machine in the front lobby.
- The bank's month-end application is applying interest to the accounts.

While your report is running, therefore, the data is being modified by multiple users, so what numbers should appear on the report? The answer depends somewhat on how your server handles *locking*, which is the mechanism used to control simultaneous use of data resources. Most database servers use one of two locking strategies:

- Database writers must request and receive from the server a *write lock* to modify data, and database readers must request and receive from the server a *read lock* to query data. While multiple users can read data simultaneously, only one write lock is given out at a time for each table (or portion thereof), and read requests are blocked until the write lock is released.

- Database writers must request and receive from the server a write lock to modify data, but readers do not need any type of lock to query data. Instead, the server ensures that a reader sees a consistent view of the data (the data seems the same even though other users may be making modifications) from the time their query begins until their query has finished. This approach is known as *versioning*.

There are pros and cons to both approaches. The first approach can lead to long wait times if there are many concurrent read and write requests, and the second approach can be problematic if there are long-running queries while data is being modified. Of the three servers discussed in this book, Microsoft SQL Server uses the first approach, Oracle Database uses the second approach, and MySQL uses both approaches (depending on your choice of *storage engine*, which is discussed a bit later).

There are also a number of different strategies that may be employed when deciding how to lock a resource. There are three different levels, or *granularities*, of locks:

Table locks
Keep multiple users from modifying any data in the same table simultaneously

Page locks
Keep multiple users from modifying data on the same page (a page is a segment of memory generally in the range of 2 KB to 16 KB) of a table simultaneously

Row locks
Keep multiple users from modifying the same row in a table simultaneously

Again, there are pros and cons to these approaches. It takes very little bookkeeping to lock entire tables, but this approach quickly yields unacceptable wait times as the number of users increases. On the other hand, row locking takes quite a bit more bookkeeping, but it allows many users to modify the same table as long as they are interested in different rows. Of the three servers discussed in this book, Microsoft SQL Server uses page and row locking, Oracle Database uses row locking, and MySQL uses table, page, or row locking (depending, again, on your choice of storage engine).

To get back to your report, the data that appears on the pages of the report will mirror either the state of the database when your report started (if your server uses a versioning approach) or the state of the database when the server issues the reporting application a read lock (if your server uses both read and write locks).

What Is a Transaction?

If database servers enjoyed 100% uptime, if users always allowed programs to finish executing, and if applications always completed without encountering fatal errors that halt execution, then there would be nothing left to discuss regarding concurrent database access. However, none of these things can be relied upon, so there is one more element necessary to allow multiple users to access the same data.

This extra piece of the concurrency puzzle is the *transaction*, which is a device for grouping together multiple SQL statements such that either *all* or *none* of the statements succeed. If you attempt to transfer $500 from your savings account to your checking account, you would be a bit upset if the money were successfully withdrawn from your savings account but never made it to your checking account. Whatever the reason for the failure (the server was shut down for maintenance, the request for a page lock on the account table timed out, etc.), you want your $500 back.

To protect against this kind of error, the program that handles your transfer request would first begin a transaction, then issue the SQL statements needed to move the money from your savings to your checking account, and, if everything succeeds, end the transaction by issuing the commit command. If something unexpected happens, however, the program would issue a rollback command, which instructs the server to undo all changes made since the transaction began. The entire process might look something like the following:

```
START TRANSACTION;

/* withdraw money from first account, making sure balance is sufficient */
UPDATE account SET avail_balance = avail_balance - 500
WHERE account_id = 9988
  AND avail_balance > 500;

IF <exactly one row was updated by the previous statement> THEN
  /* deposit money into second account */
  UPDATE account SET avail_balance = avail_balance + 500
    WHERE account_id = 9989;

  IF <exactly one row was updated by the previous statement> THEN
    /* everything worked, make the changes permanent */
    COMMIT;
  ELSE
    /* something went wrong, undo all changes in this transaction */
    ROLLBACK;
  END IF;
ELSE
  /* insufficient funds, or error encountered during update */
  ROLLBACK;
END IF;
```

 While the previous code block may look similar to one of the procedural languages provided by the major database companies, such as Oracle's PL/SQL or Microsoft's Transact SQL, it is written in pseudocode and does not attempt to mimic any particular language.

The previous code block begins by starting a transaction, and then attempts to remove $500 from the checking account and then add $500 to the savings account. If all goes well, the transaction is committed; if anything goes awry, however, then the transaction is rolled back, meaning that all data changes since the beginning of the transaction are undone.

By using a transaction, the program ensures that your five hundred dollars either stays in your savings account or moves to your checking account, without the possibility of it falling into a crack. Regardless of whether the transaction was committed or was rolled back, all resources acquired (e.g., write locks) during the execution of the transaction are released when the transaction completes.

Of course, if the program manages to complete both update statements but the server shuts down before a commit or rollback can be executed, then the transaction will be rolled back when the server comes back online. (One of the tasks that a database server must complete before coming online is to find any incomplete transactions that were underway when the server shut down and roll them back.)

Starting a Transaction

Database servers handle transaction creation in one of two ways:

- There is always an active transaction associated with a database session, so there is no need or method to explicitly begin a transaction. When a transaction ends, the server automatically begins a new transaction for your session.
- Unless you explicitly begin a transaction, individual SQL statements are automatically committed independently of one another. To begin a transaction, you must first issue a command.

Of the three servers, Oracle Database takes the first approach, while Microsoft SQL Server and MySQL take the second approach. One of the advantages of Oracle's approach to transactions is that, even if you are only issuing a single SQL command, you have the ability to roll back if you don't like the outcome or if you change your mind. Thus, if you forget to add a where clause to your delete statement, you will have the opportunity to undo the damage (assuming you've had your morning coffee and realize that you didn't mean to delete all 125,000 rows in your table). With MySQL and SQL Server, however, once you hit the Enter key, the changes brought about by your SQL statement will be permanent (unless your DBA can retrieve the original data from a backup or from some other means).

The SQL:2003 standard includes a start transaction command to be used when you want to explicitly begin a transaction. While MySQL conforms to the standard, SQL Server users must instead issue the command begin transaction. With both servers, until you explicitly begin a transaction, you are in what is known as *autocommit mode*, which means that individual statements are automatically committed by the server. You can, therefore, decide that you want to be in a transaction and issue a start/begin transaction command, or you can simply let the server commit individual statements.

Both MySQL and SQL Server allow you to turn off autocommit mode for individual sessions, in which case the servers will act just like Oracle Database regarding transactions. With SQL Server, you issue the following command to disable autocommit mode:

```
SET IMPLICIT_TRANSACTIONS ON
```

MySQL allows you to disable autocommit mode via the following:

```
SET AUTOCOMMIT=0
```

Once you have left autocommit mode, all SQL commands take place within the scope of a transaction and must be explicitly committed or rolled back.

 A word of advice: shut off autocommit mode each time you log in and get in the habit of running all of your SQL statements within a transaction. If nothing else, it may save you the embarrassment of having to ask your DBA to reconstruct data that you have inadvertently deleted.

Ending a Transaction

Once a transaction has begun, whether explicitly via the start transaction command or implicitly by the database server, you must explicitly end your transaction for your changes to become permanent. This is done by way of the commit command, which instructs the server to mark the changes as permanent and release any resources (i.e., page or row locks) used during the transaction.

If you decide that you want to undo all of the changes made since starting the transaction, you must issue the rollback command, which instructs the server to return the data to its pretransaction state. After the rollback has been completed, any resources used by your session are released.

Along with issuing either the commit or rollback command, there are several other scenarios by which your transaction can end, either as an indirect result of your actions or as a result of something outside your control:

- The server shuts down, in which case your transaction will be rolled back automatically when the server is restarted.

- You issue an SQL schema statement, such as alter table, which will cause the current transaction to be committed and a new transaction to be started.

- You issue another start transaction command, which will cause the previous transaction to be committed.

- The server prematurely ends your transaction because the server detects a *deadlock* and decides that your transaction is the culprit. In this case, the transaction will be rolled back and you will receive an error message.

Of these four scenarios, the first and third are fairly straightforward, but the other two merit some discussion. As far as the second scenario is concerned, alterations to a database, whether it be the addition of a new table or index or the removal of a column from a table, cannot be rolled back, so commands that alter your schema must take place outside of a transaction. If a transaction is currently underway, therefore, the server will commit your current transaction, execute the SQL schema statement command(s), and then automatically start a new transaction for your session. The server will not inform you of what has happened, so you should be careful that the statements that comprise a unit of work are not inadvertently broken up into multiple transactions by the server.

The fourth scenario deals with deadlock detection. A deadlock occurs when two different transactions are waiting for resources that the other transaction currently holds. For example, transaction A might have just updated the account table and is waiting for a write lock on the transaction table, while transaction B has inserted a row into the transaction table and is waiting for a write lock on the account table. If both transactions happen to be modifying the same page or row (depending on the lock granularity in use by the database server), then they will each wait forever for the other transaction to finish and free up the needed resource. Database servers must always be on the lookout for these situations so that throughput doesn't grind to a halt; when a deadlock is detected, one of the transactions is chosen (either arbitrarily or by some criteria) to be rolled back so that the other transaction may proceed.

Unlike the second scenario discussed earlier, the database server will raise an error to inform you that your transaction has been rolled back due to deadlock detection. With MySQL, for example, you will receive error #1213, which carries the following message:

```
Message: Deadlock found when trying to get lock; try restarting transaction
```

As suggested by the error message, it is a reasonable practice to retry a transaction that has been rolled back due to deadlock detection. However, if deadlocks become fairly common, then the applications that access the database may need to be modified to decrease the probability of deadlocks (one common strategy is to ensure that data resources are always accessed in the same order, such as always modifying account data before inserting transaction data).

Transaction Savepoints

In some cases, you may encounter an issue when within a transaction that requires a rollback, but you may not want to undo *all* of the work that has transpired. For these situations, you can establish one or more *savepoints* within a transaction and use them to roll back to a particular location within your transaction rather than rolling all the way back to the start of the transaction.

Choosing a Storage Engine

When using Oracle Database or Microsoft SQL Server, there is a single set of code responsible for low-level database operations, such as retrieving a particular row from a table based on primary-key value. The MySQL server, however, has been designed so that multiple storage engines may be utilized to provide low-level database functionality, including resource locking and transaction management. As of Version 4.1, MySQL includes the following storage engines:

MyISAM
> A nontransactional engine employing table locking

MEMORY
> A nontransactional engine used for in-memory tables

BDB
> A transactional engine employing page-level locking

InnoDB
> A transactional engine employing row-level locking

Merge
> A specialty engine used to make multiple identical MyISAM tables appear as a single table (a.k.a. table partitioning)

NDB
> A specialty engine used to spread a single database over multiple computers (a.k.a. clustering)

Archive
> A specialty engine used to store large amounts of unindexed data, mainly for archival purposes

While you might think that you would be forced to choose a single storage engine for your database, MySQL is flexible enough to allow you to choose a storage engine on a table-by-table basis. For any tables that might take part in transactions, however, you should choose the InnoDB storage engine, which uses row-level locking and versioning to provide the highest level of concurrency across the different storage engines.

You may explicitly specify a storage engine when creating a table, or you can change an existing table to use a different engine. If you do not know what engine is assigned to a table, you can use the show table command, as demonstrated by the following:

```
mysql> SHOW TABLE STATUS LIKE 'transaction' \G
******************** 1. row ********************
          Name: transaction
        Engine: InnoDB
  ...
 Create_options:
       Comment: InnoDB free: 3072 kB; ...
```

—continued—

> Looking at the second item, you can see that the transaction table is already using the
> InnoDB engine. If it were not, you could assign the InnoDB engine to the transaction
> table via the following command:
>
> ```
> ALTER TABLE transaction ENGINE = INNODB;
> ```

All savepoints must be given a name, which allows you to have multiple savepoints
within a single transaction. To create a savepoint named my_savepoint, you can do
the following:

```
SAVEPOINT my_savepoint;
```

To roll back to a particular savepoint, you simply issue the rollback command followed by the keywords to savepoint and the name of the savepoint, as in:

```
ROLLBACK TO SAVEPOINT my_savepoint;
```

Here's an example of how savepoints may be used:

```
START TRANSACTION;

UPDATE product
SET date_retired = CURRENT_TIMESTAMP( )
WHERE product_cd = 'XYZ';

SAVEPOINT before_close_accounts;

UPDATE account
SET status = 'CLOSED', close_date = CURRENT_TIMESTAMP( ),
  last_activity_date = CURRENT_TIMESTAMP( )
WHERE product_cd = 'XYZ';

ROLLBACK TO SAVEPOINT before_close_accounts;

COMMIT;
```

The net effect of this transaction is that the mythical XYZ product is retired but none
of the accounts are closed.

When using savepoints, remember the following:

- Despite the name, nothing is saved when you create a savepoint. You must even-
  tually issue a commit if you want your transaction to be made permanent.
- If you issue a rollback without naming a savepoint, all savepoints within the
  transaction will be ignored and the entire transaction will be undone.

If you are using SQL Server, you will need to use the proprietary command save
transaction to create a savepoint and rollback transaction to roll back to a savepoint, with each command being followed by the savepoint name.

CHAPTER 13

Indexes and Constraints

Because the focus of this book is on programming techniques, the first twelve chapters of this book concentrate on those elements of the SQL language used to craft powerful select, insert, update, and delete statements. There are, however, other database features that *indirectly* affect the code you write. This chapter focuses on two of those features: indexes and constraints.

Indexes

When you insert a row into a table, the database server does not attempt to put the data in any particular location within the table. For example, if you add a row to the department table, the server doesn't place the row in numeric order via the dept_id column or in alphabetical order via the name column. Instead, the server simply places the data in the next available location within the file (the server maintains a list of free space for each table). When you query the department table, therefore, the server will need to inspect every row of the table to answer the query. For example, let's say that you issue the following query:

```
mysql> SELECT dept_id, name
    -> FROM department
    -> WHERE name LIKE 'A%';
+---------+----------------+
| dept_id | name           |
+---------+----------------+
|       3 | Administration |
+---------+----------------+
1 row in set (0.03 sec)
```

To find all departments whose name begins with 'A', the server must visit each row in the department table and inspect the contents of the name column; if the department name begins with 'A', then the row is added to the result set.

While this method works fine for a table with only three rows, imagine how long it might take to answer the query if the table contains 3,000,000 rows. At some number of rows larger then three and smaller than 3,000,000, a line is crossed at which point the server cannot answer the query within a reasonable amount of time without additional help. This help comes in the form of one or more *indexes* on the department table.

Even if you have never heard of a database index, you are certainly aware of what an index is (for example, this book has one). An index is simply a mechanism for finding a specific item within a resource. Each technical publication, for example, includes an index at the end that allows you to locate a specific word or phrase within the publication. The index lists these words and phrases in alphabetical order, allowing the reader to move quickly to a particular letter within the index, find the desired entry, and then find the page or pages on which the word or phrase may be found.

In the same way that a person uses an index to find words within a publication, a database server uses indexes to locate rows in a table. Indexes are special tables that, unlike normal data tables, *are* kept in a specific order. Instead of containing *all* of the data about an entity, however, an index contains only the column (or columns) used to locate rows in the data table, along with information describing where the rows are physically located. Therefore, the role of indexes is to facilitate the retrieval of a subset of a table's rows and columns *without* the need to inspect every row in the table.

Index Creation

Returning to the department table, you might decide to add an index on the name column to speed up any queries that specify a full or partial department name, as well as any update or delete operations that specify a department name. Here's how you can add such an index to a MySQL database:

```
mysql> ALTER TABLE department
    -> ADD INDEX dept_name_idx (name);
Query OK, 3 rows affected (0.08 sec)
Records: 3  Duplicates: 0  Warnings: 0
```

This statement creates an index (a B-tree index to be precise, but more on this shortly) on the department.name column; furthermore, the index is given the name dept_name_idx. With the index in place, the query optimizer (which was discussed in Chapter 3) can choose to use the index if it is deemed beneficial to do so (with only three rows in the department table, for example, the optimizer might very well choose to ignore the index and read the entire table). If there is more than one index on a table, the optimizer must decide which index will be the most beneficial for a particular SQL statement.

MySQL treats indexes as optional components of a table, which is why you must use the alter table command to add or remove an index. Other database servers, including SQL Server and Oracle Database, treat indexes as independent schema objects. For both SQL Server and Oracle, therefore, you would generate an index using the create index command, as in:

```
CREATE INDEX dept_name_idx
ON department (name);
```

All database servers allow you to look at the available indexes. MySQL users can use the show command to see all of the indexes on a specific table, as in:

```
mysql> SHOW INDEX FROM department \G
*************************** 1. row ***************************
        Table: department
   Non_unique: 0
     Key_name: PRIMARY
 Seq_in_index: 1
  Column_name: dept_id
    Collation: A
  Cardinality: 3
     Sub_part: NULL
       Packed: NULL
         Null:
   Index_type: BTREE
      Comment:
*************************** 2. row ***************************
        Table: department
   Non_unique: 0
     Key_name: dept_name_uidx
 Seq_in_index: 1
  Column_name: name
    Collation: A
  Cardinality: 3
     Sub_part: NULL
       Packed: NULL
         Null:
   Index_type: BTREE
      Comment:
2 rows in set (0.02 sec)
```

The output shows that there are two indexes on the department table: one on the dept_id column called PRIMARY, and the other on the name column called dept_name_idx. Since I have only created one index so far (dept_name_idx), you might be wondering where the other came from; when the department table was created, the create table statement included a constraint naming the dept_id column as the primary key for the table. Here's the statement used to create the table:

```
CREATE TABLE department
 (dept_id SMALLINT UNSIGNED NOT NULL AUTO_INCREMENT,
  name VARCHAR(20) NOT NULL,
  CONSTRAINT pk_department PRIMARY KEY (dept_id)
 );
```

When the table was created, the MySQL server automatically generated an index on the primary-key column, which, in this case, is dept_id, and gave the index the name PRIMARY. Constraints will be covered later in this chapter.

If, after creating an index, you decide that the index is not proving useful, you can remove it via the following:

```
mysql> ALTER TABLE department
    -> DROP INDEX dept_name_idx;
Query OK, 3 rows affected (0.02 sec)
Records: 3  Duplicates: 0  Warnings: 0
```

 SQL Server and Oracle Database users must use the drop index command to remove an index, as in:

```
DROP INDEX dept_name_idx;
```

Unique indexes

When designing a database, it is important to consider which columns are allowed to contain duplicate data and which are not. For example, it is allowable to have two customers named John Smith in the individual table since each row will have a different identifier (cust_id), birth date, and tax number (customer.fed_id) to help tell them apart. You would not, however, want to allow two departments with the same name in the department table. You can enforce a rule against duplicate department names by creating a *unique index* on the department.name column.

A unique index plays multiple roles in that, along with providing all the benefits of a regular index, it also serves as a mechanism for disallowing duplicate values in the indexed column. Whenever a row is inserted or when the indexed column is modified, the database server checks the unique index to see whether the value already exists in another row in the table. Here's how you would create a unique index on the department.name column:

```
mysql> ALTER TABLE department
    -> ADD UNIQUE dept_name_idx (name);
Query OK, 3 rows affected (0.04 sec)
Records: 3  Duplicates: 0  Warnings: 0
```

 SQL Server and Oracle Database need only add the unique keyword when creating an index, as in:

```
CREATE UNIQUE INDEX dept_name_idx
ON department (name);
```

With the index in place, you will receive an error if you try to add another department with the name 'Operations':

```
mysql> INSERT INTO department (dept_id, name)
    -> VALUES (999, 'Operations');
ERROR 1062 (23000): Duplicate entry 'Operations' for key 2
```

You should not build unique indexes on your primary-key column(s), since the server already checks uniqueness for primary-key values. You may, however, create more than one unique index on the same table if you feel that it is warranted.

Multi-column indexes

Along with the single-column indexes demonstrated thus far, you may build indexes that span multiple columns. If, for example, you find yourself searching for employees by first and last names, you can build an index on *both* columns together, as in:

```
mysql> ALTER TABLE employee
    -> ADD INDEX emp_names_idx (lname, fname);
Query OK, 18 rows affected (0.10 sec)
Records: 18  Duplicates: 0  Warnings: 0
```

This index will be useful for queries that specify both the first and last names or just the last name, but it cannot be used for queries that specify only the employee's first name. To understand why, consider how you would find a person's phone number; if you know the person's first and last name, you can use a phone book to find the number quickly, since a phone book is organized by last name then by first name. If you only know the person's first name, you would need to scan every entry in the phone book to find all the entries with the specified first name.

When building multiple-column indexes, therefore, you should think carefully about which column to list first, which column to list second, etc., so that the index is a useful as possible. Keep in mind, however, that there is nothing stopping you from building multiple indexes using the same set of columns but in a different order if you feel that it is needed to ensure adequate response time.

Types of Indexes

Indexing is a powerful tool, but, since there are many different types of data, a single indexing strategy doesn't always do the job. The following sections illustrate the different types of indexing available from various servers.

B-tree indexes

All of the indexes shown thus far are *balanced-tree indexes*, which are more commonly known as *B-tree indexes*. MySQL, Oracle Database, and SQL Server all default to B-tree indexing, so you will get a B-tree index unless you explicitly ask for another type. As you might expect, B-tree indexes are organized as trees, with one or more levels of *branch nodes* leading to a single level of *leaf nodes*. Branch nodes are used for navigating the tree, while leaf nodes hold the actual values and location information. For example, a B-tree index built on the employee.lname column might look something like Figure 13-1.

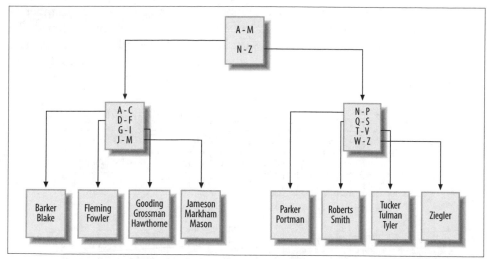

Figure 13-1. B-tree example

If you were to issue a query to retrieve all employees whose last name starts with 'G', the server would look at the top branch node (called the *root node*) and follow the link to the branch node that handles last names beginning with A through M. This branch node would, in turn, direct the server to a leaf node containing last names beginning with G through I. The server then starts reading the values in the leaf node until it encounters a value that doesn't begin with 'G' (which, in this case, is 'Hawthorne').

As rows are inserted, updated, and deleted from the employee table, the server will attempt to keep the tree balanced so that there aren't far more branch/leaf nodes on one side of the root node than the other. The server can add or remove branch nodes to redistribute the values more evenly and can even add or remove an entire level of branch nodes. By keeping the tree balanced, the server is able to traverse quickly to the leaf nodes to find the desired values without having to navigate through many levels of branch nodes.

Bitmap indexes

Whereas B-tree indexes are great at handling columns that contain many different values, such as a customer's first/last names, they can become unwieldy when built on a column that allows only a small number of values. For example, you may decide to generate an index on the account.product_cd column so that you can quickly retrieve all accounts of a specific type (e.g., checking, savings). Because there are only eight different products, however, and because some products are far more popular than others, it can be difficult to maintain a balanced B-tree index as the number of accounts grows.

For columns that contain only a small number of values across a large number of rows (known as *low-cardinality* data), a different indexing strategy is needed. To handle this situation more efficiently, Oracle Database includes *bitmap indexes*, which generate a bitmap for each value stored in the column. Figure 13-2 shows what a bitmap index might look like for data in the account.product_cd column.

Value/row	1	2	3	4	5	6	7	8	9	10	11	12	13	14	15	16	17	18	19	20	21	22	23	24
BUS	0	0	0	0	0	0	0	0	0	0	0	0	0	0	0	0	0	0	0	0	1	1	0	0
CD	0	0	1	0	0	0	0	0	0	0	0	0	1	1	0	0	0	0	1	0	0	0	0	0
CHK	1	0	0	1	0	1	0	1	0	0	1	1	0	0	1	0	1	0	0	1	0	0	1	0
MM	0	0	0	0	0	0	1	0	0	1	0	0	0	0	0	0	0	1	0	0	0	0	0	0
SAV	0	1	0	0	1	0	0	0	1	0	0	0	0	0	0	1	0	0	0	0	0	0	0	0
SBL	0	0	0	0	0	0	0	0	0	0	0	0	0	0	0	0	0	0	0	0	0	0	0	1

Figure 13-2. Bitmap example

The index contains six bitmaps, one for each value in the product_cd column (two of the eight available products are not being used), and each bitmap includes a 0/1 value for each of the 24 rows in the account table. Thus, if you ask the server to retrieve all money market accounts (product_cd = 'MM'), the server simply finds all the 1 values in the MM bitmap and returns rows 7, 10, and 18. The server can also combine bitmaps if you are looking for multiple values; for example, if you want to retrieve all money market *and* savings accounts (product_cd = 'MM' or product_cd = 'SAV'), the server can perform an OR operation on the MM and SAV bitmaps and return rows 2, 5, 7, 9, 10, 16, and 18.

Bitmap indexes are a nice, compact indexing solution for low-cardinality data, but this indexing strategy breaks down if the number of values stored in the column climbs too high in relation to the number of rows (known as *high-cardinality* data), since the server would need to maintain too many bitmaps. For example, you would never build a bitmap index on your primary-key column, since this represents the highest possible cardinality (a different value for every row).

Oracle users can generate bitmap indexes by simply adding the bitmap keyword to the create index statement, as in:

```
CREATE BITMAP INDEX acc_prod_idx ON account (product_cd);
```

Bitmap indexes are commonly used in data warehousing environments, where large amounts of data are generally indexed on columns containing relatively few values (e.g., sales quarters, geographic regions, products, salesmen).

Text indexes

If your database stores documents, you may need to allow the users to search for words or phrases in the documents. You certainly don't want the server to open each document and scan for the desired text each time a search is requested, but traditional indexing strategies don't work for this situation. To handle this situation, both MySQL and Oracle Database include specialized indexing and search mechanisms for documents; both SQL Server and MySQL include what they call *full-text indexes* (for MySQL, full-text indexes are only available with its MyISAM storage engine), and Oracle Database includes a powerful set of tools known as *Oracle Text*. Document searches are specialized enough that I will refrain from showing an example, but I wanted you to at least know what is available.

How Indexes Are Used

Indexes are generally used by the server to quickly locate rows of interest in a particular table, after which the server visits the associated table to extract the additional information requested by the user. Consider the following query:

```
mysql> SELECT emp_id, fname, lname
    -> FROM employee
    -> WHERE emp_id IN (1, 3, 9, 15, 22);
+--------+---------+----------+
| emp_id | fname   | lname    |
+--------+---------+----------+
|      1 | Michael | Smith    |
|      3 | Robert  | Tyler    |
|      9 | Jane    | Grossman |
|     15 | Frank   | Portman  |
+--------+---------+----------+
4 rows in set (0.00 sec)
```

For this query, the server can use the primary-key index on the emp_id column to locate employee IDs 1, 3, 9, 15, and 22 in the employee table, and then visit each of the five rows to retrieve the first and last name columns.

If the index contains everything needed to satisfy the query, however, then the server doesn't need to visit the associated table. To illustrate, let's look at how the query optimizer approaches the same query with different indexes in place.

The query, which aggregates account balances for specific customers, looks as follows:

```
mysql> SELECT cust_id, SUM(avail_balance) tot_bal
    -> FROM account
    -> WHERE cust_id IN (1, 5, 9, 11)
    -> GROUP BY cust_id;
+---------+----------+
| cust_id | tot_bal  |
+---------+----------+
|       1 | 4557.75  |
|       5 | 2237.97  |
```

```
|     9 | 10971.22 |
|    11 |  9345.55 |
+---------+----------+
4 rows in set (0.00 sec)
```

To see how MySQL's query optimizer decides to execute the query, I will use the explain statement to ask the server to show the execution plan for the query rather than executing the query:

```
mysql> EXPLAIN SELECT cust_id, SUM(avail_balance) tot_bal
    -> FROM account
    -> WHERE cust_id IN (1, 5, 9, 11)
    -> GROUP BY cust_id \G
*************************** 1. row ***************************
           id: 1
  select_type: SIMPLE
        table: account
         type: index
possible_keys: fk_a_cust_id
          key: fk_a_cust_id
      key_len: 4
          ref: NULL
         rows: 24
        Extra: Using where
1 row in set (0.00 sec)
```

 Each database server includes tools to allow you to see how the query optimizer handles your SQL statement. SQL Server allows you to see an execution plan by issuing the statement set showplan_text on before running your SQL statement. Oracle Database includes the explain plan statement, which writes the execution plan to a special table called plan_table.

Without going into too much detail, here's what the execution plan tells you:

- The fk_a_cust_id index is used to find the rows in the account table that satisfy the where clause.

- After reading the index, the server expects to read all 24 rows of the account table to gather the available balance data, since it doesn't know that there might be other customers besides IDs 1, 5, 9, and 11.

The fk_a_cust_id index is another index generated automatically by the server, but this time it is because of a foreign-key constraint rather than a primary-key constraint (more on this later in the chapter). The fk_a_cust_id index is on the account. cust_id column, so the server is using the index to locate customer IDs 1, 5, 9, and 11 in the account table and is then visiting those rows to retrieve and aggregate the available balance data.

Next, I will add a new index called acc_bal_idx on both the cust_id and avail_balance columns:

```
mysql> ALTER TABLE account
    -> ADD INDEX acc_bal_idx (cust_id, avail_balance);
Query OK, 24 rows affected (0.03 sec)
Records: 24  Duplicates: 0  Warnings: 0
```

With this index in place, let's see how the query optimizer approaches the same query:

```
mysql> EXPLAIN SELECT cust_id, SUM(avail_balance) tot_bal
    -> FROM account
    -> WHERE cust_id IN (1, 5, 9, 11)
    -> GROUP BY cust_id \G
*************************** 1. row ***************************
           id: 1
  select_type: SIMPLE
        table: account
         type: range
possible_keys: acc_bal_idx
          key: acc_bal_idx
      key_len: 4
          ref: NULL
         rows: 8
        Extra: Using where; Using index
1 row in set (0.01 sec)
```

Comparing the two execution plans yields the following differences:

- The optimizer is using the new acc_bal_idx index instead of the fk_a_cust_id index.

- The optimizer anticipates needing only eight rows instead of 24.

- The account table is not needed (designated by Using index in the Extra column) to satisfy the query results.

Therefore, the server can use indexes to help locate rows in the associated table, or the server can use an index as if it were a table as long as the index contains all of the columns needed by the query.

The process that I just led you through is an example of query tuning. Tuning involves looking at an SQL statement and determining the resources available to the server to execute the statement. You can decide to modify the SQL statement, adjust the database resources, or both to make a statement run more efficiently. Tuning is a detailed topic, and I strongly urge you either read your server's tuning guide or pick up a good tuning book so that you can see all of the different tuning approaches available for your server.

The Downside of Indexes

If indexes are so great, why not index everything? Well, the key to understanding why more indexes are not necessarily a good thing is to keep in mind that every index is a table (a special type of table, but still a table). Therefore, every time a row is added to or removed from a table, all indexes on that table must be modified. When a row is updated, any indexes on the column or columns that were affected need to be modified as well. Therefore, the more indexes you have, the more work the server needs to do to keep all schema objects up to date, which tends to slow things down.

Indexes also require disk space as well as some amount of care from your administrators, so the best strategy is to add an index when a clear need arises. If you only have need for an index for special purposes, such as a monthly maintenance routine, you can always add the index, run the routine, and then drop the index until you need it again. In the case of data warehouses, where indexes are crucial during business hours as users run reports and ad-hoc queries but are problematic when data is being loaded into the warehouse overnight, it is common practice to drop the indexes before data is loaded and then recreate them before the warehouse opens for business.

Constraints

A constraint is simply a restriction placed on one or more columns of a table. There are several different types of constraints, including:

Primary-key constraints
> Identify the column or columns that guarantee uniqueness within a table

Foreign-key constraints
> Restrict one or more columns to contain only values found in another table's primary-key columns, and may also restrict the allowable values in other tables if update cascade or delete cascade rules are established

Unique constraints
> Restrict one or more columns to contain unique values within a table

Check constraints
> Restrict the allowable values for a column

Without constraints, a database's consistency is suspect. For example, if the server allows you to change a customer's ID in the customer table without changing the same customer ID in the account table, then you will end up with accounts that no longer point to valid customer records (known as orphaned rows). With primary and foreign-key constraints in place, however, the server will either raise an error if an attempt is made to modify or delete data that is referenced by other tables or will propagate the changes to other tables for you (more on this shortly).

If you want to use foreign-key constraints with the MySQL server, you must use the InnoDB storage engine for your tables.

Constraint Creation

Constraints are generally created at the same time as the associated table via the create table statement. To illustrate, here's an example from the schema generation script for this book's example database:

```
CREATE TABLE product
 (product_cd VARCHAR(10) NOT NULL,
  name VARCHAR(50) NOT NULL,
  product_type_cd VARCHAR (10) NOT NULL,
  date_offered DATE,
  date_retired DATE,
  CONSTRAINT fk_product_type_cd FOREIGN KEY (product_type_cd)
    REFERENCES product_type (product_type_cd),
  CONSTRAINT pk_product PRIMARY KEY (product_cd)
 );
```

The product table includes two constraints: one to specify that the product_cd column serves as the primary key for the table, and another to specify that the product_type_cd column serves as a foreign key to the product_type table. Alternatively, the product table can be created without constraints, and the primary and foreign-key constraints can be added later via alter table statements:

```
ALTER TABLE product
ADD CONSTRAINT pk_product PRIMARY KEY (product_cd);

ALTER TABLE product
ADD CONSTRAINT fk_product_type_cd FOREIGN KEY (product_type_cd)
  REFERENCES product_type (product_type_cd);
```

If you want to remove a primary or foreign-key constraint, you can use the alter table statement again, except that you specify drop instead of add, as in:

```
ALTER TABLE product
DROP PRIMARY KEY;

ALTER TABLE product
DROP FOREIGN KEY fk_product_type_cd;
```

While it is unusual to drop a primary-key constraint, foreign-key constraints are sometimes dropped during certain maintenance operations and then reestablished.

Constraints and Indexes

As you have seen earlier in the chapter, constraint creation sometimes involves the automatic generation of an index. However, database servers behave differently regarding the relationship between constraints and indexes; Table 13-1 shows how

MySQL, SQL Server, and Oracle Database handle the relationship between constraints and indexes.

Table 13-1. Constraint generation

Constraint type	MySQL	SQL Server	Oracle Database
Primary-key constraints	Generates unique index	Generates unique index	Uses existing index or creates new index
Foreign-key constraints	Generates index	Does not generate index	Does not generate index
Unique constraints	Generates unique index	Generates unique index	Uses existing index or creates new index

MySQL, therefore, generates a new index to enforce primary-key, foreign-key, and unique constraints, SQL Server generates a new index for primary-key and unique constraints but *not* for foreign-key constraints, and Oracle Database takes the same approach as SQL Server except that Oracle will use an existing index (if an appropriate one exists) to enforce primary-key and unique constraints. While neither SQL Server or Oracle Database generate an index for a foreign-key constraint, both servers' documentation advise that indexes be created for every foreign key.

Cascading Constraints

With foreign-key constraints in place, if a user attempts to insert a new row or change an existing row such that a foreign-key column doesn't have a matching value in the parent table, the server raises an error. To illustrate, here's a look at the data in the product and product_type tables:

```
mysql> SELECT product_type_cd, name
    -> FROM product_type;
+-----------------+------------------------------+
| product_type_cd | name                         |
+-----------------+------------------------------+
| ACCOUNT         | Customer Accounts            |
| INSURANCE       | Insurance Offerings          |
| LOAN            | Individual and Business Loans |
+-----------------+------------------------------+
3 rows in set (0.00 sec)

mysql> SELECT product_type_cd, product_cd, name
    -> FROM product
    -> ORDER BY product_type_cd;
+-----------------+------------+-----------------------+
| product_type_cd | product_cd | name                  |
+-----------------+------------+-----------------------+
| ACCOUNT         | CD         | certificate of deposit |
| ACCOUNT         | CHK        | checking account      |
| ACCOUNT         | MM         | money market account  |
| ACCOUNT         | SAV        | savings account       |
| LOAN            | AUT        | auto loan             |
```

```
| LOAN            | BUS        | business line of credit |
| LOAN            | MRT        | home mortgage           |
| LOAN            | SBL        | small business loan     |
+-----------------+------------+-------------------------+
8 rows in set (0.01 sec)
```

There are three different values for the product_type_cd column in the product_type table (ACCOUNT, INSURANCE, and LOAN). Of the three values, two of them (ACCOUNT and LOAN) are referenced in the product table's product_type_cd column.

The following statement attempts to change the product_type_cd column in the product table to a value that doesn't exist in the product_type table:

```
mysql> UPDATE product
    -> SET product_type_cd = 'XYZ'
    -> WHERE product_type_cd = 'LOAN';
ERROR 1216 (23000): Cannot add or update a child row: a foreign key constraint fails
```

Because of the foreign-key constraint on the product.product_type_cd column, the server does not allow the update to succeed, since there is no row in the product_type table with a value of XYZ in the product_type_cd column. Thus, the foreign-key constraint doesn't let you change a child row if there is no corresponding value in the parent.

What would happen, however, if you tried to change the *parent* row in the product_type table to XYZ? Here's an update statement that attempts to change the LOAN product type to XYZ:

```
mysql> UPDATE product_type
    -> SET product_type_cd = 'XYZ'
    -> WHERE product_type_cd = 'LOAN';
ERROR 1217 (23000): Cannot delete or update a parent row: a foreign key constraint fails
```

Once again, an error is raised, this time because there are child rows in the product table whose product_type_cd column contains the value LOAN. This is the default behavior for foreign-key constraints, but it is not the only possible behavior; instead, you can instruct the server to propagate the change to all child rows for you, thus preserving the integrity of the data. Known as a *cascading update*, this variation of the foreign-key constraint can be installed by removing the existing foreign key and adding a new one that includes the on update cascade clause:

```
mysql> ALTER TABLE product
    -> DROP FOREIGN KEY fk_product_type_cd;
Query OK, 8 rows affected (0.02 sec)
Records: 8  Duplicates: 0  Warnings: 0

mysql> ALTER TABLE product
    -> ADD CONSTRAINT fk_product_type_cd FOREIGN KEY (product_type_cd)
    ->     REFERENCES product_type (product_type_cd)
    ->     ON UPDATE CASCADE;
Query OK, 8 rows affected (0.03 sec)
Records: 8  Duplicates: 0  Warnings: 0
```

With this modified constraint in place, let's see what happens when the previous update statement is attempted again:

```
mysql> UPDATE product_type
    -> SET product_type_cd = 'XYZ'
    -> WHERE product_type_cd = 'LOAN';
Query OK, 1 row affected (0.01 sec)
Rows matched: 1  Changed: 1  Warnings: 0
```

This time, the statement succeeds. To verify that the change was propagated to the product table, here's another look at the data in both tables:

```
mysql> SELECT product_type_cd, name
    -> FROM product_type;
+-----------------+------------------------------+
| product_type_cd | name                         |
+-----------------+------------------------------+
| ACCOUNT         | Customer Accounts            |
| INSURANCE       | Insurance Offerings          |
| XYZ             | Individual and Business Loans |
+-----------------+------------------------------+
3 rows in set (0.02 sec)

mysql> SELECT product_type_cd, product_cd, name
    -> FROM product
    -> ORDER BY product_type_cd;
+-----------------+------------+------------------------+
| product_type_cd | product_cd | name                   |
+-----------------+------------+------------------------+
| ACCOUNT         | CD         | certificate of deposit |
| ACCOUNT         | CHK        | checking account       |
| ACCOUNT         | MM         | money market account   |
| ACCOUNT         | SAV        | savings account        |
| XYZ             | AUT        | auto loan              |
| XYZ             | BUS        | business line of credit |
| XYZ             | MRT        | home mortgage          |
| XYZ             | SBL        | small business loan    |
+-----------------+------------+------------------------+
8 rows in set (0.01 sec)
```

As you can see, the change to the product_type table has been propagated to the product table as well. Along with cascading updates, you can specify *cascading deletes* as well. A cascading delete removes rows from the child table when a row is deleted from the parent table. To specify cascading deletes, use the on delete cascade clause, as in:

```
ALTER TABLE product
ADD CONSTRAINT fk_product_type_cd FOREIGN KEY (product_type_cd)
  REFERENCES product_type (product_type_cd)
  ON UPDATE CASCADE
  ON DELETE CASCADE;
```

With this version of the constraint in place, the server will now update child rows in the product table when a row in the product_type table is updated as well as delete child rows in the product table when a row in the product_type table is deleted.

Cascading constraints are one case in which constraints *do* directly affect the code that you write. You need to know which constraints in your database specify cascading updates and/or deletes so that you know the full effect of your update and delete statements.

ER Diagram for Example Database

Figure A-1 is an entity-relationship (ER) diagram for the example database used in this book. As the name suggests, the diagram depicts the entities, or tables, in the database along with the foreign-key relationships between the tables. Here are a few tips to help you understand the notation:

- Each rectangle represents a table, with the table name above the upper-left corner of the rectangle. The primary-key column(s) are listed first and are separated from nonkey columns by a line. Nonkey columns are listed below the line, and foreign key columns are marked with "(FK)."

- Lines between tables represent foreign key relationships. The markings at either end of the lines represents the allowable quantity, which can be zero (0), one (1), or many (≤). For example, if you look at the relationship between the account and product tables, you would say that an account must belong to exactly one product, but a product may have zero, one, or many accounts.

For more information on database modeling and modeling tools, see Appendix D.

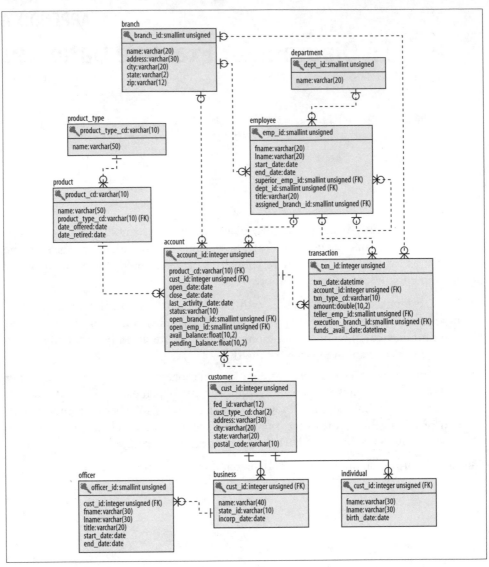

Figure A-1. ER diagram

MySQL Extensions to the SQL Language

Since this book uses the MySQL server for all of the examples, I thought it would be useful for those readers planning to continue using MySQL to include an appendix on MySQL's extensions to the SQL language. This appendix explores some of MySQL's extensions to the select, insert, update, and delete statements that can be very useful in certain situations.

Select Extensions

MySQL's implementation of the select statement includes two additional clauses, which are discussed in the following subsections.

The limit Clause

In some situations, you may not be interested in *all* of the rows returned by a query. For example, you might construct a query that returns all of the bank tellers along with the number of accounts opened by each teller. If your reason for executing the query is to determine the top three tellers so that they can receive an award from the bank, then you don't necessarily need to know who came in fourth, fifth, etc. To help with these types of situations, MySQL's select statement includes the limit clause, which allows you to restrict the number of rows returned by a query.

To demonstrate the utility of the limit clause, I will begin by constructing a query to show the number of accounts opened by each bank teller:

```
mysql> SELECT open_emp_id, COUNT(*) how_many
    -> FROM account
    -> GROUP BY open_emp_id;
+-------------+----------+
| open_emp_id | how_many |
+-------------+----------+
|           1 |        8 |
|          10 |        7 |
|          13 |        3 |
```

```
|          16 |        6 |
+-------------+----------+
4 rows in set (0.31 sec)
```

The results show that four different tellers opened accounts; if you want to limit the result set to only three records, you can add a limit clause specifying that only three records be returned:

```
mysql> SELECT open_emp_id, COUNT(*) how_many
    -> FROM account
    -> GROUP BY open_emp_id
    -> LIMIT 3;
+-------------+----------+
| open_emp_id | how_many |
+-------------+----------+
|           1 |        8 |
|          10 |        7 |
|          13 |        3 |
+-------------+----------+
3 rows in set (0.06 sec)
```

Thanks to the limit clause (the fourth line of the query), the result set now includes exactly three records, and the fourth teller (employee ID 16) has been discarded from the result set.

Combining the limit clause with the order by clause

While the previous query returns three records, there's one small problem; you haven't described *which* three of the four records you are interested in. If you are looking for three *specific* records, such as the three tellers who opened the most accounts, you will need to use the limit clause in concert with an order by clause, as in:

```
mysql> SELECT open_emp_id, COUNT(*) how_many
    -> FROM account
    -> GROUP BY open_emp_id
    -> ORDER BY how_many DESC
    -> LIMIT 3;
+-------------+----------+
| open_emp_id | how_many |
+-------------+----------+
|           1 |        8 |
|          10 |        7 |
|          16 |        6 |
+-------------+----------+
3 rows in set (0.03 sec)
```

The difference between this query and the previous query is that the limit clause is now being applied to an ordered set, resulting in the three tellers with the most opened accounts being included in the final result set. Unless you are only interested in seeing an arbitrary sample of records, you will generally want to use an order by clause along with a limit clause.

The limit clause is applied after all filtering, grouping, and ordering has occurred, so it will never change the outcome of your select statement other than restricting the number of records returned by the statement.

limit's optional second parameter

Instead of finding the top three tellers, let's say your goal is to identify all but the top two tellers (instead of giving awards to top performers, the bank will be sending some of the less-productive tellers to assertiveness training). For these types of situations, the limit clause allows for an optional second parameter; when two parameters are used, the first designates at which record to begin adding records to the final result set, and the second designates how many records to include. When specifying a record by number, remember that MySQL designates the first record as record 0. Therefore, if your goal is to find the third-best performer, you can do the following:

```
mysql> SELECT open_emp_id, COUNT(*) how_many
    -> FROM account
    -> GROUP BY open_emp_id
    -> ORDER BY how_many DESC
    -> LIMIT 2, 1;
+-------------+----------+
| open_emp_id | how_many |
+-------------+----------+
|          16 |        6 |
+-------------+----------+
1 row in set (0.00 sec)
```

In this example, the zeroth and first records are discarded, and records are included starting at the second record. Since the second parameter in the limit clause is a 1, only a single record is included.

If you want to start at the second position and include *all* of the remaining records, you can make the second argument to the limit clause large enough to guarantee that all remaining records are included. If you do not know how many tellers opened new accounts, therefore, you might do something like the following to find all but the top two performers:

```
mysql> SELECT open_emp_id, COUNT(*) how_many
    -> FROM account
    -> GROUP BY open_emp_id
    -> ORDER BY how_many DESC
    -> LIMIT 2, 999999999;
+-------------+----------+
| open_emp_id | how_many |
+-------------+----------+
|          16 |        6 |
|          13 |        3 |
+-------------+----------+
2 rows in set (0.00 sec)
```

In this version of the query, the zeroth and first records are discarded, and up to 999,999,999 records are included starting at the second record (in this case, there are only two more, but it's better to go a bit overboard rather than taking a chance on excluding valid records from your final result set because you underestimated).

Ranking queries

When used in conjunction with an order by clause, queries that include a limit clause can be called *ranking queries* because they allow you to rank your data. While I have demonstrated how to rank bank tellers by the number of opened accounts, ranking queries are used to answer many different types of business questions, such as:

- Who are the top five salespeople for 2005?
- Who has the third-most home runs in the history of baseball?
- Other than *The Holy Bible* and *Quotations from Chairman Mao*, what are the next 98 best-selling books of all time?
- What are our two worst-selling flavors of ice cream?

So far, I have shown how to find the top three tellers, the third-best teller, and all but the top two tellers. If I want to do something analogous to the fourth example (i.e., find the *worst* performers), I need only reverse the sort order so that the results proceed from lowest number of accounts opened to highest number of accounts opened, as in:

```
mysql> SELECT open_emp_id, COUNT(*) how_many
    -> FROM account
    -> GROUP BY open_emp_id
    -> ORDER BY how_many ASC
    -> LIMIT 2;
+-------------+----------+
| open_emp_id | how_many |
+-------------+----------+
|          13 |        3 |
|          16 |        6 |
+-------------+----------+
2 rows in set (0.24 sec)
```

By simply changing the sort order (from ORDER BY how_many DESC to ORDER BY how_many ASC), the query now returns the two worst-performing tellers. Therefore, by using a limit clause with either an ascending or descending sort order, you can produce ranking queries to answer most types of business questions.

The into outfile Clause

If you want the output from your query to be written to a file, you could highlight the query results, copy them to the buffer, and paste them into your favorite editor. However, if the query's result set is sufficiently large, or if the query is being executed from within a script, you will need a way to write the results to a file without your intervention. To aid in such situations, MySQL includes the into outfile clause to allow you to provide the name of a file into which the results will be written. Here's an example that writes the query results to a file in my *c:\temp* directory:

```
mysql> SELECT emp_id, fname, lname, start_date
    -> INTO OUTFILE 'C:\\TEMP\\emp_list.txt'
    -> FROM employee;
Query OK, 18 rows affected (0.20 sec)
```

If you remember from Chapter 7, the backslash is used to escape another character within a string. If you're a Windows user, therefore, you will need to put two backslashes in a row when building path names.

Rather than showing the query results on the screen, the result set has been written to the *emp_list.txt* file, which looks as follows:

```
1    Michael    Smith     2001-06-22
2    Susan      Barker    2002-09-12
3    Robert     Tyler     2000-02-09
4    Susan      Hawthorne    2002-04-24
...
16    Theresa    Markham    2001-03-15
17    Beth    Fowler    2002-06-29
18    Rick    Tulman    2002-12-12
```

The default format uses tabs ('\t') between columns and newlines ('\n') after each record. If you want more control over the format of the data, there are several additional subclauses available with the into outfile clause. For example, if you want the data to be in what is referred to as *pipe-delimited format*, you can use the fields subclause to ask that the '|' character be placed between each column, as in:

```
mysql> SELECT emp_id, fname, lname, start_date
    -> INTO OUTFILE 'C:\\TEMP\\emp_list_delim.txt'
    ->     FIELDS TERMINATED BY '|'
    -> FROM employee;
Query OK, 18 rows affected (0.02 sec)
```

MySQL does not allow you to overwrite an existing file when using into outfile, so you will need to remove an existing file first if you run the same query more than once.

The contents of the *emp_list_delim.txt* file look as follows:

```
1|Michael|Smith|2001-06-22
2|Susan|Barker|2002-09-12
3|Robert|Tyler|2000-02-09
4|Susan|Hawthorne|2002-04-24
...
16|Theresa|Markham|2001-03-15
17|Beth|Fowler|2002-06-29
18|Rick|Tulman|2002-12-12
```

Along with pipe-delimited format, you may need your data in *comma-delimited format*, in which case you would use fields terminated by ','. If the data being written to a file includes strings, however, using commas as field separators can prove problematic, since commas are much more likely to appear within strings than the pipe character. Consider the following query, which writes a number and two strings delimited by commas to the *comma1.txt* file:

```
mysql> SELECT data.num, data.str1, data.str2
    -> INTO OUTFILE 'C:\\TEMP\\comma1.txt'
    ->    FIELDS TERMINATED BY ','
    -> FROM
    -> (SELECT 1 num, 'This string has no commas' str1,
    ->      'This string, however, has two commas' str2) data;
Query OK, 1 row affected (0.04 sec)
```

Since the third column in the output file (str2) is a string containing commas, you might think that an application attempting to read the *comma1.txt* file will encounter problems when parsing each line into columns, but the MySQL server has made provisions for such situations. Here are the contents of *comma1.txt*:

```
1,This string has no commas,This string\, however\, has two commas
```

As you can see, the commas within the third column have been escaped by putting a backslash before the two commas embedded in the str2 column. If you run the same query but use pipe-delimited format, the commas will *not* be escaped, since they don't need to be. If you want to use a different escape character, such as using another comma, you can use the fields escaped by subclause to specify the escape character to use for your output file.

Along with specifying column separators, you can also specify the character used to separate the different records in your data file. If you would like each record in the output file to be separated by something other than the newline character, you can use the lines subclause, as in:

```
mysql> SELECT emp_id, fname, lname, start_date
    -> INTO OUTFILE 'C:\\TEMP\\emp_list_atsign.txt'
    ->    FIELDS TERMINATED BY '|'
    ->    LINES TERMINATED BY '@'
    -> FROM employee;
Query OK, 18 rows affected (0.03 sec)
```

Because I am not using a newline character between records, the *emp_list_atsign.txt* file looks like a single long line of text when viewed, with each record separated by the '@' character:

```
1|Michael|Smith|2001-06-22@2|Susan|Barker|2002-09-12@3|Robert|Tyler|2000-02-
09@4|Susan|Hawthorne|2002-04-24@5|John|Gooding|2003-11-14@6|Helen|Fleming|2004-03-
17@7|Chris|Tucker|2004-09-15@8|Sarah|Parker|2002-12-02@9|Jane|Grossman|2002-05-
03@10|Paula|Roberts|2002-07-27@11|Thomas|Ziegler|2000-10-23@12|Samantha|Jameson|2003-
01-08@13|John|Blake|2000-05-11@14|Cindy|Mason|2002-08-09@15|Frank|Portman|2003-04-
01@16|Theresa|Markham|2001-03-15@17|Beth|Fowler|2002-06-29@18|Rick|Tulman|2002-12-12@
```

If you need to generate a data file to be loaded into a spreadsheet application or sent within or outside your organization, the into outfile clause should provide enough flexibility for whatever file format you need.

Combination Insert/Update Statements

Let's say that you have been asked to create a table to capture information about which of the bank's branches are visited by which customers. The table needs to contain the customer's ID, the branch's ID, and a datetime column indicating the last time the customer visited the branch. Rows are added to the table whenever a customer visits a certain branch, but if the customer has already visited the branch then the existing row should simply have its datetime column updated. Here's the table definition:

```
CREATE TABLE branch_usage
 (branch_id SMALLINT UNSIGNED NOT NULL,
  cust_id INTEGER UNSIGNED NOT NULL,
  last_visited_on DATETIME,
  CONSTRAINT pk_branch_usage PRIMARY KEY (branch_id, cust_id)
 );
```

Along with the three column definitions, the branch_usage table defines a primary-key constraint on the branch_id and cust_id columns. Therefore, any row added to the table whose branch/customer pair already exists in the table will be rejected by the server.

Let's say that, after the table is in place, customer ID 5 visits the main branch (branch ID 1) three times in the first week. After the first visit, you can insert a record into the branch_usage table, since no record exists yet for customer ID 5 and branch ID 1:

```
mysql> INSERT INTO branch_usage (branch_id, cust_id, last_visited_on)
    -> VALUES (1, 5, CURRENT_TIMESTAMP());
Query OK, 1 row affected (0.02 sec)
```

The next time the customer visits the same branch, however, you will need to *update* the existing record rather than inserting a new record; otherwise, you will receive the following error:

```
ERROR 1062 (23000): Duplicate entry '1-5' for key 1
```

To avoid this error, you can query the branch_usage table to see whether a given customer/branch pair exists and then either insert a record if no record is found or update the existing row if it already exists. To save you the trouble, however, the MySQL designers have extended the insert statement to allow you to specify that one or more columns be modified if an insert statement fails due to a duplicate key. The following statement instructs the server to modify the last_visited_on column if the given customer and branch already exist in the branch_usage table:

```
mysql> INSERT INTO branch_usage (branch_id, cust_id, last_visited_on)
    -> VALUES (1, 5, CURRENT_TIMESTAMP())
```

```
-> ON DUPLICATE KEY UPDATE last_visited_on = CURRENT_TIMESTAMP( );
Query OK, 2 rows affected (0.02 sec)
```

The on duplicate key clause allows this same statement to be executed every time customer ID 5 conducts business in branch ID 1. If run 100 times, the first execution results in a single row being added to the table, and the next 99 executions result in the last_visited_on column being changed to the current time. This type of operation is often referred to as an *upsert*, since it is a combination of an update and an insert statement.

Replacing the replace Command

Prior to Version 4.1 of the MySQL server, upsert operations were performed using the replace command, which is a proprietary statement that first deletes an existing row if the primary-key value already exists in the table before inserting a row. If you are using Version 4.1 or later, you can choose between the replace command and the insert... on duplicate key command when performing upsert operations.

However, the replace command performs a delete operation when duplicate key values are encountered, which can cause a ripple effect if you are using the InnoDB storage engine and have foreign-key constraints enabled. If the constraints have been created with the on delete cascade option, then rows in other tables may also be automatically deleted when the replace command deletes a row in the target table. For this reason, it is generally regarded as safer to use the on duplicate key clause of the insert statement rather than the older replace command.

Ordered Updates and Deletes

Earlier in the chapter, I showed you how write queries using the limit clause in conjunction with an order by clause to generate rankings, such as the top three tellers in terms of accounts opened. MySQL also allows the limit and order by clauses to be used in both update and delete statements, thereby allowing you to modify or remove specific rows in a table based on a ranking. For example, imagine that you are asked to remove records from a table used to track customer logins to the bank's online banking system. The table, which tracks the customer ID and date/time of login, looks as follows:

```
CREATE TABLE login_history
 (cust_id INTEGER UNSIGNED NOT NULL,
  login_date DATETIME,
  CONSTRAINT pk_login_history PRIMARY KEY (cust_id, login_date)
 );
```

The following statement populates the login_history table with some data by generating a cross join between the account and customer tables and using the account's open_date column as a basis for generating login dates:

```
mysql> INSERT INTO login_history (cust_id, login_date)
    -> SELECT c.cust_id,
    ->   ADDDATE(a.open_date, INTERVAL a.account_id * c.cust_id HOUR)
    -> FROM customer c CROSS JOIN account a;
Query OK, 312 rows affected (0.03 sec)
Records: 312  Duplicates: 0  Warnings: 0
```

The table is now populated with 312 rows of relatively random data. Your task is to look at the data in the login_history table once a month, generate a report for your manager showing who is using the online banking system, and then to delete all but the 50 most-recent records from the table. One approach would be to write a query using order by and limit to find the 50th most recent login, such as:

```
mysql> SELECT login_date
    -> FROM login_history
    -> ORDER BY login_date DESC
    -> LIMIT 49,1;
+---------------------+
| login_date          |
+---------------------+
| 2004-07-02 09:00:00 |
+---------------------+
1 row in set (0.00 sec)
```

Armed with this information, you can then construct a delete statement that removes all rows whose login_date column is less than the date returned by the query:

```
mysql> DELETE FROM login_history
    -> WHERE login_date < '2004-07-02 09:00:00';
Query OK, 262 rows affected (0.02 sec)
```

The table now contains the 50 most-recent logins. Using MySQL's extensions, however, the same result can be achieved with a single delete statement using limit and order by clauses. After returning the original 312 rows to the login_history table, you can run the following:

```
mysql> DELETE FROM login_history
    -> ORDER BY login_date ASC
    -> LIMIT 262;
Query OK, 262 rows affected (0.05 sec)
```

With this statement, the rows are sorted by login_date in ascending order, and then the first 262 rows are deleted, leaving the 50 most recent rows.

In this example, I had to know the number of rows in the table to construct the limit clause (312 original rows − 50 remaining rows = 262 deletions). It would be better if you could sort the rows in descending order and tell the server to skip the first 50 rows and then delete the remaining rows, as in:

```
DELETE FROM login_history
ORDER BY login_date DESC
LIMIT 49, 9999999;
```

However, MySQL does not allow the optional second parameter when using the limit clause in delete or update statements.

Along with deleting data, you can use the limit and order by clauses when modifying data as well. For example, if the bank decides to add $100 to each of the ten oldest accounts to help retain loyal customers, you can do the following:

```
mysql> UPDATE account
    -> SET avail_balance = avail_balance + 100
    -> WHERE product_cd IN ('CHK', 'SAV', 'MM')
    -> ORDER BY open_date ASC
    -> LIMIT 10;
Query OK, 10 rows affected (0.06 sec)
Rows matched: 10  Changed: 10  Warnings: 0
```

This statement sorts accounts by the open date in ascending order and then modifies the first ten records, which, in this case, are the ten oldest accounts.

Multitable Updates and Deletes

In certain situations, you might need to modify or delete data from several different tables to perform a given task. If you discover that the bank's database contains a dummy customer left over from system testing, for example, you might need to remove data from the account, customer, and individual tables.

For this section, I will create a set of clones for the account, customer, and individual tables, called account2, customer2, and individual2. I am doing so both to protect the sample data from being altered and to avoid any problems with foreign-key constraints between the tables (more on this later in the section). Here's the create table statements used to generate the three clone tables:

```
CREATE TABLE individual2 AS
SELECT * FROM individual;

CREATE TABLE customer2 AS
SELECT * FROM customer;

CREATE TABLE account2 AS
SELECT * FROM account;
```

If the customer ID of the dummy customer is 1, you could generate three individual delete statements against each of the three tables, as in:

```
DELETE FROM account2
WHERE cust_id = 1;

DELFTE FROM customer2
WHERE cust_id = 1;

DELETE FROM individual2
WHERE cust_id = 1;
```

Instead of writing individual delete statements, however, MySQL allows you to write a single *multitable* delete statement, which, in this case, looks as follows:

```
mysql> DELETE account2, customer2, individual2
    -> FROM account2 INNER JOIN customer2
    ->   ON account2.cust_id = customer2.cust_id
    ->   INNER JOIN individual2
    ->   ON customer2.cust_id = individual2.cust_id
    -> WHERE individual2.cust_id = 1;
Query OK, 5 rows affected (0.02 sec)
```

This statement removes a total of five rows, one from each of the individual2 and customer2 tables, and three from the account2 table (customer ID 1 has three accounts). The statement comprises three separate clauses:

delete

Specifies the tables targeted for deletion.

from

Specifies the tables used to identify the rows to be deleted. This clause is identical in form and function to the from clause in a select statement, and not all tables named herein need to be included in the delete clause.

where

Contains filter conditions used to identify the rows to be deleted.

The multitable delete statement looks a lot like a select statement, except that a delete clause is used instead of a select clause. If you are deleting rows from a single table using a multitable delete format, the difference becomes even less noticeable. For example, here's a select statement that finds the account IDs of all accounts owned by John Hayward:

```
mysql> SELECT account2.account_id
    -> FROM account2 INNER JOIN customer2
    ->   ON account2.cust_id = customer2.cust_id
    ->   INNER JOIN individual2
    ->   ON individual2.cust_id = customer2.cust_id
    -> WHERE individual2.fname = 'John'
    ->   AND individual2.lname = 'Hayward';
+------------+
| account_id |
+------------+
```

```
|           8 |
|           9 |
|          10 |
+-------------+
3 rows in set (0.01 sec)
```

If, after viewing the results, you decide to delete all three of John's accounts from the account2 table, you need only replace the select clause in the previous query with a delete clause naming the account2 table, as in:

```
mysql> DELETE account2
    -> FROM account2 INNER JOIN customer2
    ->   ON account2.cust_id = customer2.cust_id
    ->   INNER JOIN individual2
    ->   ON customer2.cust_id = individual2.cust_id
    -> WHERE individual2.fname = 'John'
    ->   AND individual2.lname = 'Hayward';
Query OK, 3 rows affected (0.01 sec)
```

Hopefully, this gives you a better idea of what the delete and from clauses are used for in a multitable delete statement. This statement is functionally identical to the following single-table delete statement, which uses a subquery to identify the customer ID of John Hayward:

```
DELETE FROM account2
WHERE cust_id =
 (SELECT cust_id
  FROM individual2
  WHERE fname = 'John' AND lname = 'Hayward';
```

When using a multitable delete statement to delete rows from a single table, you are simply choosing to use a query-like format involving table joins rather than a traditional delete statement using subqueries. The real power of multitable delete statements lies in the ability to delete from multiple tables in a single statement, as was demonstrated in the first statement in this section.

Along with the ability to delete rows from multiple tables, MySQL also gives you the ability to *modify* rows in multiple tables using a *multitable update*. Let's say that your bank is merging with another bank, and the databases from both banks have overlapping customer IDs. Your management decides to fix the problem by adding 10,000 to each customer ID in your database so that the second bank's data can be safely imported. The following statement shows how to modify the ID of customer ID 3 across the individual2, customer2, and account2 tables using a single statement:

```
mysql> UPDATE individual2 INNER JOIN customer2
    ->   ON individual2.cust_id = customer2.cust_id
    ->   INNER JOIN account2
    ->   ON customer2.cust_id = account2.cust_id
    -> SET individual2.cust_id = individual2.cust_id + 10000,
    ->   customer2.cust_id = customer2.cust_id + 10000,
    ->   account2.cust_id = account2.cust_id + 10000
    -> WHERE individual2.cust_id = 3;
```

```
Query OK, 4 rows affected (0.01 sec)
Rows matched: 5  Changed: 4  Warnings: 0
```

This statement modifies four rows: one in each of the individual2 and customer2 tables, and two in the account2 table. The multitable update syntax is very similar to that of the single-table update, except that the update clause contains multiple tables and their corresponding join conditions rather than just naming a single table. Just like the single-table update, the multitable version includes a set clause, the difference being that any tables referenced in the update clause may be modified via the set clause.

 If you are using the InnoDB storage engine, you will most likely not be able to use multitable delete and update statements if the tables involved have foreign-key constraints. This is because the engine does not guarantee that the changes will be applied in an order that won't violate the constraints. Instead, you should use multiple single-table statements in the proper order so that foreign key constraints are not violated.

Solutions to Exercises

Chapter 3

3-1

Retrieve the employee ID, first name, and last name for all bank employees. Sort by last name then first name.

```
mysql> SELECT emp_id, fname, lname
    -> FROM employee
    -> ORDER BY lname, fname;
+--------+----------+-----------+
| emp_id | fname    | lname     |
+--------+----------+-----------+
|      2 | Susan    | Barker    |
|     13 | John     | Blake     |
|      6 | Helen    | Fleming   |
|     17 | Beth     | Fowler    |
|      5 | John     | Gooding   |
|      9 | Jane     | Grossman  |
|      4 | Susan    | Hawthorne |
|     12 | Samantha | Jameson   |
|     16 | Theresa  | Markham   |
|     14 | Cindy    | Mason     |
|      8 | Sarah    | Parker    |
|     15 | Frank    | Portman   |
|     10 | Paula    | Roberts   |
|      1 | Michael  | Smith     |
|      7 | Chris    | Tucker    |
|     18 | Rick     | Tulman    |
|      3 | Robert   | Tyler     |
|     11 | Thomas   | Ziegler   |
+--------+----------+-----------+
18 rows in set (0.01 sec)
```

3-2

Retrieve the account ID, customer ID, and available balance for all accounts whose status equals 'ACTIVE' and whose available balance is greater than $2,500.

```
mysql> SELECT account_id, cust_id, avail_balance
    -> FROM account
    -> WHERE status = 'ACTIVE'
    ->   AND avail_balance > 2500;
+------------+---------+---------------+
| account_id | cust_id | avail_balance |
+------------+---------+---------------+
|          3 |       1 |       3000.00 |
|         10 |       4 |       5487.09 |
|         13 |       6 |      10000.00 |
|         14 |       7 |       5000.00 |
|         15 |       8 |       3487.19 |
|         18 |       9 |       9345.55 |
|         20 |      10 |      23575.12 |
|         22 |      11 |       9345.55 |
|         23 |      12 |      38552.05 |
|         24 |      13 |      50000.00 |
+------------+---------+---------------+
10 rows in set (0.00 sec)
```

3-3

Write a query against the account table that returns the IDs of the employees who opened the accounts (use the account.open_emp_id column). Include a single row for each distinct employee.

```
mysql> SELECT DISTINCT open_emp_id
    -> FROM account;
+-------------+
| open_emp_id |
+-------------+
|           1 |
|          10 |
|          13 |
|          16 |
+-------------+
4 rows in set (0.00 sec)
```

3-4

Fill in the blanks (denoted by <#>) for this multi-data-set query to achieve the results shown.

```
mysql> SELECT p.product_cd, a.cust_id, a.avail_balance
    -> FROM product p INNER JOIN account <1>
    ->   ON p.product_cd = <2>
    -> WHERE p.<3> = 'ACCOUNT';
```

```
+------------+---------+---------------+
| product_cd | cust_id | avail_balance |
+------------+---------+---------------+
| CD         |       1 |       3000.00 |
| CD         |       6 |      10000.00 |
| CD         |       7 |       5000.00 |
| CD         |       9 |       1500.00 |
| CHK        |       1 |       1057.75 |
| CHK        |       2 |       2258.02 |
| CHK        |       3 |       1057.75 |
| CHK        |       4 |        534.12 |
| CHK        |       5 |       2237.97 |
| CHK        |       6 |        122.37 |
| CHK        |       8 |       3487.19 |
| CHK        |       9 |        125.67 |
| CHK        |      10 |      23575.12 |
| CHK        |      12 |      38552.05 |
| MM         |       3 |       2212.50 |
| MM         |       4 |       5487.09 |
| MM         |       9 |       9345.55 |
| SAV        |       1 |        500.00 |
| SAV        |       2 |        200.00 |
| SAV        |       4 |        767.77 |
| SAV        |       8 |        387.99 |
+------------+---------+---------------+
21 rows in set (0.02 sec)
```

The correct values for <1>, <2>, and <3> are:

1. a
2. a.product_cd
3. product_type_cd

Chapter 4

4-1

Which of the transaction IDs would be returned by the following filter conditions?

```
txn_date < '2005-02-26' AND (txn_type_cd = 'DBT' OR amount > 100)
```

Transaction IDs 1, 2, 3, 5, 6, and 7.

4-2

Which of the transaction IDs would be returned by the following filter conditions?

```
account_id IN (101,103) AND NOT (txn_type_cd = 'DBT' OR amount > 100)
```

Transaction IDs 4 and 9.

4-3

Construct a query that retrieves all accounts opened in 2002.

```
mysql> SELECT account_id, open_date
    -> FROM account
    -> WHERE open_date BETWEEN '2002-01-01' AND '2002-12-31';
+------------+------------+
| account_id | open_date  |
+------------+------------+
|          6 | 2002-11-23 |
|          7 | 2002-12-15 |
|         12 | 2002-08-24 |
|         20 | 2002-09-30 |
|         21 | 2002-10-01 |
+------------+------------+
5 rows in set (0.01 sec)
```

4-4

Construct a query that finds all nonbusiness customers whose last name contains an 'a' in the second position and an 'e' anywhere after the 'a'.

```
mysql> SELECT cust_id, lname, fname
    -> FROM individual
    -> WHERE lname LIKE '_a%e%';
+---------+--------+---------+
| cust_id | lname  | fname   |
+---------+--------+---------+
|       1 | Hadley | James   |
|       9 | Farley | Richard |
+---------+--------+---------+
2 rows in set (0.02 sec)
```

Chapter 5

5-1

Fill in the blanks (denoted by <#>) for the following query to obtain the results that follow:

```
mysql> SELECT e.emp_id, e.fname, e.lname, b.name
    -> FROM employee e INNER JOIN <1> b
    ->   ON e.assigned_branch_id = b.<2>;
+--------+---------+-----------+--------------+
| emp_id | fname   | lname     | name         |
+--------+---------+-----------+--------------+
|      1 | Michael | Smith     | Headquarters |
|      2 | Susan   | Barker    | Headquarters |
|      3 | Robert  | Tyler     | Headquarters |
|      4 | Susan   | Hawthorne | Headquarters |
|      5 | John    | Gooding   | Headquarters |
```

```
|      6 | Helen    | Fleming  | Headquarters  |
|      7 | Chris    | Tucker   | Headquarters  |
|      8 | Sarah    | Parker   | Headquarters  |
|      9 | Jane     | Grossman | Headquarters  |
|     10 | Paula    | Roberts  | Woburn Branch |
|     11 | Thomas   | Ziegler  | Woburn Branch |
|     12 | Samantha | Jameson  | Woburn Branch |
|     13 | John     | Blake    | Quincy Branch |
|     14 | Cindy    | Mason    | Quincy Branch |
|     15 | Frank    | Portman  | Quincy Branch |
|     16 | Theresa  | Markham  | So. NH Branch |
|     17 | Beth     | Fowler   | So. NH Branch |
|     18 | Rick     | Tulman   | So. NH Branch |
+--------+----------+----------+---------------+
18 rows in set (0.03 sec)
```

The correct values for <1> and <2> are:

1. branch

2. branch_id

5-2

Write a query that returns the account ID for each nonbusiness customer (customer.cust_type_cd = 'I') along with the customer's federal ID (customer.fed_id) and the name of the product on which the account is based (product.name).

```
mysql> SELECT a.account_id, c.fed_id, p.name
    -> FROM account a INNER JOIN customer c
    ->   ON a.cust_id = c.cust_id
    ->   INNER JOIN product p
    ->   ON a.product_cd = p.product_cd
    -> WHERE c.cust_type_cd = 'I';
+------------+-------------+-----------------------+
| account_id | fed_id      | name                  |
+------------+-------------+-----------------------+
|          1 | 111-11-1111 | checking account      |
|          2 | 111-11-1111 | savings account       |
|          3 | 111-11-1111 | certificate of deposit |
|          4 | 222-22-2222 | checking account      |
|          5 | 222-22-2222 | savings account       |
|          6 | 333-33-3333 | checking account      |
|          7 | 333-33-3333 | money market account  |
|          8 | 444-44-4444 | checking account      |
|          9 | 444-44-4444 | savings account       |
|         10 | 444-44-4444 | money market account  |
|         11 | 555-55-5555 | checking account      |
|         12 | 666-66-6666 | checking account      |
|         13 | 666-66-6666 | certificate of deposit |
|         14 | 777-77-7777 | certificate of deposit |
|         15 | 888-88-8888 | checking account      |
|         16 | 888-88-8888 | savings account       |
|         17 | 999-99-9999 | checking account      |
```

```
|          18 | 999-99-9999 | money market account  |
|          19 | 999-99-9999 | certificate of deposit |
+-------------+-------------+------------------------+
19 rows in set (0.00 sec)
```

5-3

Construct a query that finds all employees whose supervisor is assigned to a different department. Retrieve the employees' ID, first name, and last name.

```
mysql> SELECT e.emp_id, e.fname, e.lname
    -> FROM employee e INNER JOIN employee mgr
    ->   ON e.superior_emp_id = mgr.emp_id
    -> WHERE e.dept_id != mgr.dept_id;
+--------+-------+-----------+
| emp_id | fname | lname     |
+--------+-------+-----------+
|      4 | Susan | Hawthorne |
|      5 | John  | Gooding   |
+--------+-------+-----------+
2 rows in set (0.00 sec)
```

Chapter 6

6-1

If set A = {L M N O P} and set B = {P Q R S T}, what sets are generated by the following operations:

- A union B = {L M N O P Q R S T}
- A union all B = {L M N O P P Q R S T}
- A intersect B = {P}
- A except B = {L M N O}

6-2

Write a compound query that finds the first and last names of all individual customers along with the first and last names of all employees.

```
mysql> SELECT fname, lname
    -> FROM individual
    -> UNION
    -> SELECT fname, lname
    -> FROM employee;
+----------+-----------+
| fname    | lname     |
+----------+-----------+
| James    | Hadley    |
| Susan    | Tingley   |
```

```
| Frank    | Tucker    |
| John     | Hayward   |
| Charles  | Frasier   |
| John     | Spencer   |
| Margaret | Young     |
| Louis    | Blake     |
| Richard  | Farley    |
| Michael  | Smith     |
| Susan    | Barker    |
| Robert   | Tyler     |
| Susan    | Hawthorne |
| John     | Gooding   |
| Helen    | Fleming   |
| Chris    | Tucker    |
| Sarah    | Parker    |
| Jane     | Grossman  |
| Paula    | Roberts   |
| Thomas   | Ziegler   |
| Samantha | Jameson   |
| John     | Blake     |
| Cindy    | Mason     |
| Frank    | Portman   |
| Theresa  | Markham   |
| Beth     | Fowler    |
| Rick     | Tulman    |
+----------+-----------+
27 rows in set (0.01 sec)
```

6-3

Sort the results from exercise 6-2 by the lname column.

```
mysql> SELECT fname, lname
    -> FROM individual
    -> UNION ALL
    -> SELECT fname, name
    -> FROM employee
    -> ORDER BY lname;
+----------+-----------+
| fname    | lname     |
+----------+-----------+
| Susan    | Barker    |
| Louis    | Blake     |
| John     | Blake     |
| Richard  | Farley    |
| Helen    | Fleming   |
| Beth     | Fowler    |
| Charles  | Frasier   |
| John     | Gooding   |
| Jane     | Grossman  |
| James    | Hadley    |
```

```
| Susan    | Hawthorne |
| John     | Hayward   |
| Samantha | Jameson   |
| Theresa  | Markham   |
| Cindy    | Mason     |
| Sarah    | Parker    |
| Frank    | Portman   |
| Paula    | Roberts   |
| Michael  | Smith     |
| John     | Spencer   |
| Susan    | Tingley   |
| Chris    | Tucker    |
| Frank    | Tucker    |
| Rick     | Tulman    |
| Robert   | Tyler     |
| Margaret | Young     |
| Thomas   | Ziegler   |
+----------+-----------+
27 rows in set (0.01 sec)
```

Chapter 7

7-1

Write a query that returns the 17th through 25th characters of the string "Please find the substring in this string."

```
mysql> SELECT SUBSTRING('Please find the substring in this string',17,9);
+-----------------------------------------------------------+
| SUBSTRING('Please find the substring in this string',17,9) |
+-----------------------------------------------------------+
| substring                                                 |
+-----------------------------------------------------------+
1 row in set (0.00 sec)
```

7-2

Write a query that returns the absolute value and sign (–1, 0, or 1) of the number –25.76823. Also return the number rounded to the nearest hundredth.

```
mysql> SELECT ABS(-25.76823), SIGN(-25.76823), ROUND(-25.76823, 2);
+----------------+-----------------+---------------------+
| ABS(-25.76823) | SIGN(-25.76823) | ROUND(-25.76823, 2) |
+----------------+-----------------+---------------------+
|       25.76823 |              -1 |              -25.77 |
+----------------+-----------------+---------------------+
1 row in set (0.00 sec)
```

7-3

Write a query to return just the month portion of the current date.

```
mysql> SELECT EXTRACT(MONTH FROM CURRENT_DATE());
+----------------------------------+
| EXTRACT(MONTH FROM CURRENT_DATE) |
+----------------------------------+
|                                5 |
+----------------------------------+
1 row in set (0.02 sec)
```

(Your result will most likely be different, unless it happens to be May when you try this exercise.)

Chapter 8

8-1

Construct a query that counts the number of rows in the account table.

```
mysql> SELECT COUNT(*)
    -> FROM account;
+----------+
| count(*) |
+----------+
|       24 |
+----------+
1 row in set (0.32 sec)
```

8-2

Modify your query from exercise 8-1 to count the number of accounts held by each customer. Show the customer ID and the number of accounts for each customer.

```
mysql> SELECT cust_id, COUNT(*)
    -> FROM account
    -> GROUP BY cust_id;
+---------+----------+
| cust_id | count(*) |
+---------+----------+
|       1 |        3 |
|       2 |        2 |
|       3 |        2 |
|       4 |        3 |
|       5 |        1 |
|       6 |        2 |
|       7 |        1 |
|       8 |        2 |
|       9 |        3 |
|      10 |        2 |
|      11 |        1 |
```

```
|      12 |        1 |
|      13 |        1 |
+---------+----------+
13 rows in set (0.00 sec)
```

8-3

Modify your query from exercise 8-2 to only include those customers having at least two accounts.

```
mysql> SELECT cust_id, COUNT(*)
    -> FROM account
    -> GROUP BY cust_id
    -> HAVING COUNT(*) >= 2;
+---------+----------+
| cust_id | COUNT(*) |
+---------+----------+
|       1 |        3 |
|       2 |        2 |
|       3 |        2 |
|       4 |        3 |
|       6 |        2 |
|       8 |        2 |
|       9 |        3 |
|      10 |        2 |
+---------+----------+
8 rows in set (0.04 sec)
```

8-4 (Extra Credit)

Find the total available balance by product and branch where there is more than one account per product and branch. Order the results by total balance (highest to lowest).

```
mysql> SELECT product_cd, open_branch_id, SUM(avail_balance)
    -> FROM account
    -> GROUP BY product_cd, open_branch_id
    -> HAVING COUNT(*) > 1
    -> ORDER BY 3 DESC;
+------------+----------------+--------------------+
| product_cd | open_branch_id | SUM(avail_balance) |
+------------+----------------+--------------------+
| CHK        |              4 |           67852.33 |
| MM         |              1 |           14832.64 |
| CD         |              1 |           11500.00 |
| CD         |              2 |            8000.00 |
| CHK        |              2 |            3315.77 |
| CHK        |              1 |             782.16 |
| SAV        |              2 |             700.00 |
+------------+----------------+--------------------+
7 rows in set (0.01 sec)
```

Note: MySQL would not accept ORDER BY SUM(avail_balance) DESC,, so I was forced to indicate the sort column by position.

Chapter 9

9-1

Construct a query against the account table that uses a filter condition with a noncorrelated subquery against the product table to find all loan accounts (product.product_type_cd = 'LOAN'). Retrieve the account ID, product code, customer ID, and available balance.

```
mysql> SELECT account_id, product_cd, cust_id, avail_balance
    -> FROM account
    -> WHERE product_cd IN (SELECT product_cd
    ->    FROM product
    ->    WHERE product_type_cd = 'LOAN');
+------------+------------+---------+---------------+
| account_id | product_cd | cust_id | avail_balance |
+------------+------------+---------+---------------+
|         21 | BUS        |      10 |          0.00 |
|         22 | BUS        |      11 |       9345.55 |
|         24 | SBL        |      13 |      50000.00 |
+------------+------------+---------+---------------+
3 rows in set (0.07 sec)
```

9-2

Rework the query from exercise 9-1 using a *correlated* subquery against the product table to achieve the same results.

```
mysql> SELECT a.account_id, a.product_cd, a.cust_id, a.avail_balance
    -> FROM account a
    -> WHERE EXISTS (SELECT 1
    ->    FROM product p
    ->    WHERE p.product_cd = a.product_cd
    ->      AND p.product_type_cd = 'LOAN');
+------------+------------+---------+---------------+
| account_id | product_cd | cust_id | avail_balance |
+------------+------------+---------+---------------+
|         21 | BUS        |      10 |          0.00 |
|         22 | BUS        |      11 |       9345.55 |
|         24 | SBL        |      13 |      50000.00 |
+------------+------------+---------+---------------+
3 rows in set (0.01 sec)
```

9-3

Join the following query to the employee table to show the experience level of each employee:

```
SELECT 'trainee' name, '2004-01-01' start_dt, '2005-12-31' end_dt
UNION ALL
SELECT 'worker' name, '2002-01-01' start_dt, '2003-12-31' end_dt
UNION ALL
SELECT 'mentor' name, '2000-01-01' start_dt, '2001-12-31' end_dt
```

Give the subquery the alias levels, and include the employee's ID, first name, last name, and experience level (levels.name). (Hint: Build a join condition using an inequality condition to determine into which level the employee.start_date column falls.)

```
mysql> SELECT e.emp_id, e.fname, e.lname, levels.name
    -> FROM employee e INNER JOIN
    -> (SELECT 'trainee' name, '2004-01-01' start_dt, '2005-12-31' end_dt
    -> UNION ALL
    -> SELECT 'worker' name, '2002-01-01' start_dt, '2003-12-31' end_dt
    -> UNION ALL
    -> SELECT 'mentor' name, '2000-01-01' start_dt, '2001-12-31' end_dt) levels
    -> ON e.start_date BETWEEN levels.start_dt AND levels.end_dt;
+--------+----------+-----------+---------+
| emp_id | fname    | lname     | name    |
+--------+----------+-----------+---------+
|      6 | Helen    | Fleming   | trainee |
|      7 | Chris    | Tucker    | trainee |
|      2 | Susan    | Barker    | worker  |
|      4 | Susan    | Hawthorne | worker  |
|      5 | John     | Gooding   | worker  |
|      8 | Sarah    | Parker    | worker  |
|      9 | Jane     | Grossman  | worker  |
|     10 | Paula    | Roberts   | worker  |
|     12 | Samantha | Jameson   | worker  |
|     14 | Cindy    | Mason     | worker  |
|     15 | Frank    | Portman   | worker  |
|     17 | Beth     | Fowler    | worker  |
|     18 | Rick     | Tulman    | worker  |
|      1 | Michael  | Smith     | mentor  |
|      3 | Robert   | Tyler     | mentor  |
|     11 | Thomas   | Ziegler   | mentor  |
|     13 | John     | Blake     | mentor  |
|     16 | Theresa  | Markham   | mentor  |
+--------+----------+-----------+---------+
18 rows in set (0.00 sec)
```

9-4

Construct a query against the employee table that retrieves the employee ID, first name, and last name, along with the names of the department and branch to which the employee is assigned. Do not join any tables.

```
mysql> SELECT e.emp_id, e.fname, e.lname,
    ->   (SELECT d.name FROM department d
    ->    WHERE d.dept_id = e.dept_id) dept_name,
    ->   (SELECT b.name FROM branch b
    ->    WHERE b. branch_id = e.assigned_branch_id) branch_name
    -> FROM employee e;
+--------+----------+-----------+----------------+---------------+
| emp_id | fname    | lname     | dept_name      | branch_name   |
+--------+----------+-----------+----------------+---------------+
|      1 | Michael  | Smith     | Administration | Headquarters  |
|      2 | Susan    | Barker    | Administration | Headquarters  |
|      3 | Robert   | Tyler     | Administration | Headquarters  |
|      4 | Susan    | Hawthorne | Operations     | Headquarters  |
|      5 | John     | Gooding   | Loans          | Headquarters  |
|      6 | Helen    | Fleming   | Operations     | Headquarters  |
|      7 | Chris    | Tucker    | Operations     | Headquarters  |
|      8 | Sarah    | Parker    | Operations     | Headquarters  |
|      9 | Jane     | Grossman  | Operations     | Headquarters  |
|     10 | Paula    | Roberts   | Operations     | Woburn Branch |
|     11 | Thomas   | Ziegler   | Operations     | Woburn Branch |
|     12 | Samantha | Jameson   | Operations     | Woburn Branch |
|     13 | John     | Blake     | Operations     | Quincy Branch |
|     14 | Cindy    | Mason     | Operations     | Quincy Branch |
|     15 | Frank    | Portman   | Operations     | Quincy Branch |
|     16 | Theresa  | Markham   | Operations     | So. NH Branch |
|     17 | Beth     | Fowler    | Operations     | So. NH Branch |
|     18 | Rick     | Tulman    | Operations     | So. NH Branch |
+--------+----------+-----------+----------------+---------------+
18 rows in set (0.12 sec)
```

Chapter 10

10-1

Write a query that returns all product names along with the accounts based on that product (use the product_cd column in the account table to link to the product table). Include all products, even if no accounts have been opened for that product.

```
mysql> SELECT p.product_cd, a.account_id, a.cust_id, a.avail_balance
    -> FROM product p LEFT OUTER JOIN account a
    ->   ON p.product_cd = a.product_cd;
+------------+------------+---------+---------------+
| product_cd | account_id | cust_id | avail_balance |
+------------+------------+---------+---------------+
| AUT        |       NULL |    NULL |          NULL |
| BUS        |         21 |      10 |          0.00 |
```

```
| BUS        |         22 |       11 |        9345.55 |
| CD         |          3 |        1 |        3000.00 |
| CD         |         13 |        6 |       10000.00 |
| CD         |         14 |        7 |        5000.00 |
| CD         |         19 |        9 |        1500.00 |
| CHK        |          1 |        1 |        1057.75 |
| CHK        |          4 |        2 |        2258.02 |
| CHK        |          6 |        3 |        1057.75 |
| CHK        |          8 |        4 |         534.12 |
| CHK        |         11 |        5 |        2237.97 |
| CHK        |         12 |        6 |         122.37 |
| CHK        |         15 |        8 |        3487.19 |
| CHK        |         17 |        9 |         125.67 |
| CHK        |         20 |       10 |       23575.12 |
| CHK        |         23 |       12 |       38552.05 |
| MM         |          7 |        3 |        2212.50 |
| MM         |         10 |        4 |        5487.09 |
| MM         |         18 |        9 |        9345.55 |
| MRT        |       NULL |     NULL |           NULL |
| SAV        |          2 |        1 |         500.00 |
| SAV        |          5 |        2 |         200.00 |
| SAV        |          9 |        4 |         767.77 |
| SAV        |         16 |        8 |         387.99 |
| SBL        |         24 |       13 |       50000.00 |
+------------+------------+----------+----------------+
26 rows in set (0.01 sec)
```

10-2

Reformulate your query from exercise 10-1 to use the other outer join type (i.e., if you used a left-outer join in 10-1, use a right-outer join this time) such that the results are identical to 10-1.

```
mysql> SELECT p.product_cd, a.account_id, a.cust_id, a.avail_balance
    -> FROM account a RIGHT OUTER JOIN product p
    ->   ON p.product_cd = a.product_cd;
+------------+------------+----------+----------------+
| product_cd | account_id | cust_id  | avail_balance  |
+------------+------------+----------+----------------+
| AUT        |       NULL |     NULL |           NULL |
| BUS        |         21 |       10 |           0.00 |
| BUS        |         22 |       11 |        9345.55 |
| CD         |          3 |        1 |        3000.00 |
| CD         |         13 |        6 |       10000.00 |
| CD         |         14 |        7 |        5000.00 |
| CD         |         19 |        9 |        1500.00 |
| CHK        |          1 |        1 |        1057.75 |
| CHK        |          4 |        2 |        2258.02 |
| CHK        |          6 |        3 |        1057.75 |
| CHK        |          8 |        4 |         534.12 |
| CHK        |         11 |        5 |        2237.97 |
| CHK        |         12 |        6 |         122.37 |
| CHK        |         15 |        8 |        3487.19 |
```

```
| CHK      |        17 |        9 |         125.67 |
| CHK      |        20 |       10 |       23575.12 |
| CHK      |        23 |       12 |       38552.05 |
| MM       |         7 |        3 |        2212.50 |
| MM       |        10 |        4 |        5487.09 |
| MM       |        18 |        9 |        9345.55 |
| MRT      |      NULL |     NULL |           NULL |
| SAV      |         2 |        1 |         500.00 |
| SAV      |         5 |        2 |         200.00 |
| SAV      |         9 |        4 |         767.77 |
| SAV      |        16 |        8 |         387.99 |
| SBL      |        24 |       13 |       50000.00 |
+----------+-----------+----------+----------------+
26 rows in set (0.02 sec)
```

10-3

Outer join the account table to both the individual and business tables (via the account.cust_id column) such that the result set contains one row per account. Columns to include are account.account_id, account.product_cd, individual.fname, individual.lname, and business.name.

```
mysql> SELECT a.account_id, a.product_cd,
    ->   i.fname, i.lname, b.name
    -> FROM account a LEFT OUTER JOIN business b
    ->   ON a.cust_id = b.cust_id
    ->   LEFT OUTER JOIN individual i
    ->   ON a.cust_id = i.cust_id;
+------------+------------+----------+----------+------------------------+
| account_id | product_cd | fname    | lname    | name                   |
+------------+------------+----------+----------+------------------------+
|          1 | CHK        | James    | Hadley   | NULL                   |
|          2 | SAV        | James    | Hadley   | NULL                   |
|          3 | CD         | James    | Hadley   | NULL                   |
|          4 | CHK        | Susan    | Tingley  | NULL                   |
|          5 | SAV        | Susan    | Tingley  | NULL                   |
|          6 | CHK        | Frank    | Tucker   | NULL                   |
|          7 | MM         | Frank    | Tucker   | NULL                   |
|          8 | CHK        | John     | Hayward  | NULL                   |
|          9 | SAV        | John     | Hayward  | NULL                   |
|         10 | MM         | John     | Hayward  | NULL                   |
|         11 | CHK        | Charles  | Frasier  | NULL                   |
|         12 | CHK        | John     | Spencer  | NULL                   |
|         13 | CD         | John     | Spencer  | NULL                   |
|         14 | CD         | Margaret | Young    | NULL                   |
|         15 | CHK        | Louis    | Blake    | NULL                   |
|         16 | SAV        | Louis    | Blake    | NULL                   |
|         17 | CHK        | Richard  | Farley   | NULL                   |
|         18 | MM         | Richard  | Farley   | NULL                   |
|         19 | CD         | Richard  | Farley   | NULL                   |
|         20 | CHK        | NULL     | NULL     | Chilton Engineering    |
|         21 | BUS        | NULL     | NULL     | Chilton Engineering    |
|         22 | BUS        | NULL     | NULL     | Northeast Cooling Inc. |
```

```
|          23 | CHK        | NULL      | NULL      | Superior Auto Body      |
|          24 | SBL        | NULL      | NULL      | AAA Insurance Inc.      |
+-------------+------------+-----------+---------+-------------------------+
24 rows in set (0.05 sec)
```

10-4 (Extra Credit)

Devise a query that will generate the set {1, 2, 3,..., 99, 100}. (Hint: Use a cross join with at least two from clause subqueries.)

```
SELECT ones.x + tens.x + 1
FROM
 (SELECT 0 x UNION ALL
  SELECT 1 x UNION ALL
  SELECT 2 x UNION ALL
  SELECT 3 x UNION ALL
  SELECT 4 x UNION ALL
  SELECT 5 x UNION ALL
  SELECT 6 x UNION ALL
  SELECT 7 x UNION ALL
  SELECT 8 x UNION ALL
  SELECT 9 x) ones
CROSS JOIN
 (SELECT 0 x UNION ALL
  SELECT 10 x UNION ALL
  SELECT 20 x UNION ALL
  SELECT 30 x UNION ALL
  SELECT 40 x UNION ALL
  SELECT 50 x UNION ALL
  SELECT 60 x UNION ALL
  SELECT 70 x UNION ALL
  SELECT 80 x UNION ALL
  SELECT 90 x) tens;
```

Chapter 11

11-1

Rewrite the following query, which uses a simple case expression, so that the same results are achieved using a searched case expression. Try to use as few when clauses as possible.

```
SELECT emp_id,
  CASE title
    WHEN 'President' THEN 'Management'
    WHEN 'Vice President' THEN 'Management'
    WHEN 'Treasurer' THEN 'Management'
    WHEN 'Loan Manager' THEN 'Management'
    WHEN 'Operations Manager' THEN 'Operations'
    WHEN 'Head Teller' THEN 'Operations'
    WHEN 'Teller' THEN 'Operations'
```

```
      ELSE 'Unknown'
    END
FROM employee;

SELECT emp_id,
  CASE
    WHEN title LIKE '%President' OR title = 'Loan Manager'
      OR title = 'Treasurer'
      THEN 'Management'
    WHEN title LIKE '%Teller' OR title = 'Operations Manager'
      THEN 'Operations'
    ELSE 'Unknown'
  END
FROM employee;
```

11-2

Rewrite the following query so that the result set contains a single row with four columns (one for each branch). Name the four columns branch_1 through branch_4.

```
mysql> SELECT open_branch_id, COUNT(*)
    -> FROM account
    -> GROUP BY open_branch_id;
+----------------+----------+
| open_branch_id | COUNT(*) |
+----------------+----------+
|              1 |        8 |
|              2 |        7 |
|              3 |        3 |
|              4 |        6 |
+----------------+----------+
4 rows in set (0.00 sec)

mysql> SELECT
    ->     SUM(CASE WHEN open_branch_id = 1 THEN 1 ELSE 0 END) branch_1,
    ->     SUM(CASE WHEN open_branch_id = 2 THEN 1 ELSE 0 END) branch_2,
    ->     SUM(CASE WHEN open_branch_id = 3 THEN 1 ELSE 0 END) branch_3,
    ->     SUM(CASE WHEN open_branch_id = 4 THEN 1 ELSE 0 END) branch_4
    -> FROM account;
+----------+----------+----------+----------+
| branch_1 | branch_2 | branch_3 | branch_4 |
+----------+----------+----------+----------+
|        8 |        7 |        3 |        6 |
+----------+----------+----------+----------+
1 row in set (0.02 sec)
```

Further Resources

Now that you have finished reading this book, you should be well on your way toward proficiency with the SQL language. Since I decided to be a bit more aggressive with the depth of coverage in this book versus a typical introductory book, your grasp of some of the topics might still be a bit hazy. This is a good thing, since, in my opinion, the purchase of a technical book that requires only a single read is a waste of money. I hope that you will reread certain chapters and continue to experiment with the sample database until you have a solid grasp of the concepts.

The next step to take on your journey depends on your particular goals. After having worked with many people over the years, I would guess that you probably fall into one of the following categories:

- You are a programmer (or are working on a computer science degree) with little or no prior database knowledge, and you either want to broaden your skill set or have been asked to help out with the database aspects of a project.

- You are not a programmer but have been asked to work on a reporting or Business Intelligence project at your company, possibly including the installation and administration of a BI server such as Business Objects, Actuate, Microstrategy, or Cognos.

- You are a systems administrator and want to broaden your scope to include database administration.

- You are a small business owner who needs a database to keep track of customers, inventory, orders, etc., or you are using a packaged application and want to write custom reports.

- You are none of these things but have, for whatever reason, had a database dropped in your lap (not the actual computer, hopefully).

Depending on which, if any, of these categories best describes your situation, you might be interested in investigating one or more of the following topics:

- Advanced SQL
- Database programming

- Database design
- Database tuning
- Database administration
- Report generation

The remainder of this appendix will introduce each of these topics and suggest some resources to help you master these additional skills.

Advanced SQL

No matter what role you are playing in your organization, be it administrator, programmer, performance expert, report designer, or even database designer, there is no skill more important to your personal success and the success of your project than mastery of the SQL language. That being said, it is a bit tricky to know when you have mastered a language that includes only a handful of commands (I am referring to the SQL data statements category, which includes the select, insert, update, and delete statements). The best way to determine whether you have mastered SQL is to take an advanced course or read an advanced book; afterward, most people feel that they didn't know the topic as well as they thought they did.

Here's how mastery of the SQL language can help you:

- Programmers who access databases are ultimately responsible for the proper implementation of the project specifications and for the overall performance of the system. The best programmers are able to craft concise, efficient SQL statements that properly implement the project specifications without causing performance problems.

- Database administration often encompasses several skills, including database design and implementation, database programming, and database tuning. Administrators need to master the SQL schema statements (such as create table and alter index) to perform their day-to-day administration tasks, but mastery of the SQL data statements will help them make better decisions for their design, programming, and tuning tasks.

- Performance experts generally agree that most performance problems are caused by poorly-conceived SQL statements and not by lack of server resources. It is common practice, however, to lay blame on the server when a performance problem arises and to initiate costly and unnecessary hardware upgrades.

- Good database designs must be relatively easy to navigate, or else the people implementing the design will make mistakes. The more you know about how the data will be retrieved from your database, the better decisions you will make during the design phase.

When studying advanced SQL, you will venture outside of the core language into specialized interfaces and command sets. For example, your SQL implementation may allow you to generate Extensible Markup Language (XML) directly from a database query or to store, parse, and retrieve XML documents. Your SQL implementation might also include specialized functionality for data warehouse and business intelligence queries, such as the ability to generate rankings (e.g., show me the top 10 salespeople last year). Additionally, you may need to interface with object-oriented languages and might need to make use of specialized commands for storing, retrieving, and creating objects and collections of objects. None of these topics are typically covered in introductory SQL books.

The following books are good for taking your SQL to the next level:

Mastering Oracle SQL, Second Edition
Sanjay Mishra and Alan Beaulieu
O'Reilly, 2004

MySQL Cookbook
Paul DuBois
O'Reilly, 2002

Microsoft SQL Server 2000 Bible
Paul Nielsen
Wiley, 2002

You may also consider taking a course through one of the following training centers:

* Oracle University (*http://education.oracle.com*)
* Learning Tree International (*http://www.learningtree.com*)
* Microsoft Learning (*http://www.microsoft.com/learning*)
* MySQL Training (*http://www.mysql.com/training*)

Database Programming

If you are a programmer looking to add database access to your bag of tricks, learning SQL is just one piece of the puzzle; you also need a language or API that will allow you to create database sessions and interact with the database via SQL commands. You may already work with a language that has these capabilities built in, or you may need to use an additional API or driver. Table D-1 shows some of the options available for the major programming languages.

Table D-1. Database access options

Language	API	Description
Java	Java Database Connectivity (JDBC)	A set of interfaces for database interaction. You must obtain a JDBC driver (an implementation of the JDBC interfaces) from your database provider or from a third party.
C++	Oracle Call Interface (OCI)	A set of C/C++ libraries for connecting to an Oracle database and issuing SQL commands.
	MySQL++	A set of C++ libraries for connecting to a MySQL database and issuing SQL commands.
	RogueWave SourcePro DB	A set of C++ libraries for connecting to a MySQL, Oracle, or SQL Server database (among others) and issuing SQL commands.
Perl	DBI	A module for accessing MySQL, SQL Server, Oracle, and several other database engines via a single interface.
Visual C++ Visual C# Visual Basic	Microsoft ActiveX® Data Objects .NET (ADO.NET)	A set of interfaces for providing data access services for the .NET platform.

All of the languages shown in Table D-1 are general-purpose languages, and all of them require an additional driver or set of libraries to access a database. There are other languages, however, that are designed specifically for database access and include certain SQL commands (at a minimum, select, update, insert, delete, start transaction, commit, and rollback) in the language's grammar. Table D-2 lists some of these languages and describes the environment in which they operate.

Table D-2. Database-specific languages

Language	Provider	Runtime environment
PL/SQL	Oracle	Oracle Database Oracle Application Server
Transact-SQL	Microsoft	SQL Server
SQL2003 Stored Procedure Language	MySQL	MySQL Server (Version 5.0 and up)
PowerBuilder	Sybase	Sybase EAServer Windows and Unix clients
PowerHouse 4GL	Cognos	PowerHouse Application Server
Progress 4GL	Progress Software	OpenEdge Database OpenEdge Application Server

While the last three languages shown in Table D-2 are general-purpose languages used to create business applications, the first three languages, which are provided by the three database servers covered in this book, are used to generate the following types of modules:

Stored procedures
> Named routines that accept parameters

Stored functions
> Named functions that accept parameters and return a value

Triggers
> Modules that are run automatically by the database server when a specific event occurs, such as the deletion of data in a particular table

While triggers are executed only by the database server, stored procedures and functions are executed by way of a database session just like SQL commands. Because stored functions return a value, they can be called from within SQL statements wherever a scalar subquery may be used. If you will be working with Oracle Database, SQL Server, or MySQL, you should seriously consider picking up one of the books listed below:

> *JDBC API Tutorial and Reference*, Third Edition
> Maydene Fisher et. al.
> Addison-Wesley, 2003

> *ADO.NET Cookbook*
> Bill Hamilton
> O'Reilly, 2003

> *Programming the Perl DBI*
> Alligator Descartes, Tim Bunce
> O'Reilly, 2000

> *Oracle PL/SQL Programming*, Third Edition
> Steven Feuerstein with Bill Pribyl
> O'Reilly, 2002

> *The Guru's Guide to Transact-SQL*
> Ken Henderson
> Addison-Wesley, 2000

There are also many database programming courses available through one of the following training centers:

- Oracle University (*http://education.oracle.com*)
- Learning Tree International (*http://www.learningtree.com*)

- Microsoft Learning (*http://www.microsoft.com/learning*)
- MySQL Training (*http://www.mysql.com/training*)

Database Design

If you are new to SQL (which I assume you are), you will most likely be working with existing databases, at least at first. If you are also charged with designing a database for your project, however, I advise you to take a more thorough look at database design than the brief discussion of relational database design and normalization in Chapter 2. There are actually several flavors of database design, each one having a specific purpose:

Logical models
> Typically a high-level view of an organization and the environment in which the organization conducts business

Functional models
> Typically a medium-level view of a particular segment of an organization's business, generally used as an accompaniment to a project specification

Physical models
> Typically used to generate databases

If you are a database administrator, you may only be interested in physical models, whereas logical models are often the domain of enterprise architects (if your organization is fortunate enough to have an enterprise architecture team).

However you go about it, you should seriously consider using a modeling tool to build visual models before jumping into create table statements. When building database models, you will generally be using one of the following two methodologies:

Entity-relationship (ER) modeling
> Used almost exclusively for database modeling

Unified Modeling Language (UML) modeling
> A general-purpose modeling tool for object-oriented software development

If you are designing a database as part of an object-oriented software project, then your team might buy a UML modeling tool to do object modeling and expect you to use it for your database design as well. If you are free to choose whatever tool you wish, you may find one of the following ER tools to be a bit more useful, since they can generate fully-functional database schemas (including tables, constraints, indexes, views, etc.) at the touch of a button:

- ERwin Data Modeler (ER modeling)
- Computer Associates (*http://www.ca.com*)
- ER/Studio (ER modeling)
- Embarcadero Technologies (*http://www.embarcadero.com*)

- Rational Rose (UML modeling)
- IBM (*http://www.ibm.com*)
- Visio (both ER and UML modeling)
- Microsoft (*http://www.microsoft.com*)

Following is a short list of some good books on database design:

> *Database Design for Mere Mortals: A Hands-on Guide to Relational Database Design*, Second Edition
> Michael J. Hernandez
> Addison-Wesley, 2003
>
> *UML for Database Design*
> Eric J. Naiburg and Robert A. Maksimchuk
> Addison-Wesley, 2001

Database Tuning

Database tuning is, in essence, the art and science of finding and eliminating performance bottlenecks within:

- Applications that access databases (SQL, locking, transactions)
- Database schemas (table designing, indexing, partitioning)
- Database servers (server configuration, logging, connection management)
- Disk arrays that hold database files (RAID configurations, hotspot detection)
- Computers that host database servers (operating system configuration, filesystems)
- Networks that propagate data to or from database servers

Finding and fixing bottlenecks across such a range of hardware and software might seem like a daunting job description, but most of the work is generally focused on the database schema and the SQL used by the applications that access the database. This is not meant to belittle the tasks of configuring an operating system, installing and configuring a database server, and laying out data resources on a disk array, but database schemas and the SQL used to access them are much more dynamic components of a system and are, therefore, much more apt to cause problems.

Whether you are a full-time performance expert, a database programmer, or a database administrator, most of your tuning activities will revolve around the following:

- Viewing the execution plan of SQL statements to look for inefficiencies
- Evaluating indexing strategies to ensure efficient access
- Tweaking or rewriting SQL statements to influence choice of execution plan

As I mentioned in Chapter 3, every database includes a component called a *query optimizer*, whose job it is to evaluate SQL statements and determine how to efficiently access data resources to achieve the desired results. The output of the optimizer is an *execution plan* that shows which resources are used in what order. Each of the three databases discussed in this book include tools for capturing and viewing the execution plan of an SQL statement, and you will need to learn how to generate and decipher execution plans for your database.

Just to give you a taste, here's an execution plan generated by MySQL for a query that accesses two tables:

```
mysql> EXPLAIN SELECT c.fed_id, a.account_id, a.avail_balance
    -> FROM account a INNER JOIN customer c
    ->   ON a.cust_id = c.cust_id
    -> WHERE c.cust_type_cd = 'I' \G
*************************** 1. row ***************************
           id: 1
  select_type: SIMPLE
        table: c
         type: ALL
possible_keys: PRIMARY
          key: NULL
      key_len: NULL
          ref: NULL
         rows: 13
        Extra: Using where
*************************** 2. row ***************************
           id: 1
  select_type: SIMPLE
        table: a
         type: ref
possible_keys: fk_a_cust_id
          key: fk_a_cust_id
      key_len: 4
          ref: bank.c.cust_id
         rows: 1
        Extra:
2 rows in set (0.00 sec)
```

To see the execution plan for this query, I simply preface the select statement with the explain keyword, which tells the server to show the execution plan rather than the result set for the query. The plan has two steps; the first step shows how the customer table will be accessed (all rows are accessed, since there is no index on the cust_type_cd column), and the second shows how the account table will be accessed (via the fk_a_cust_id foreign key). Learning how to generate and decipher execution plans is not a topic for an introductory book on SQL, so please see the resources at the end of this section for ideas on books and training classes. There are also some excellent tools on the market that will help you capture, evaluate, and tune SQL statements; some of these tools are listed next as well:

- Quest Central for Oracle
- Quest Central for SQL Server
- Quest Software (*http://www.quest.com*)
- Embarcadero SQL Profiler
- Embarcadero Technologies (*http://www.embarcadero.com*)
- Oracle Enterprise Manager Tuning Pack
- Oracle Corporation (*http://www.oracle.com*)

There are some very good books that I can recommend in the area of SQL performance tuning:

Effective Oracle by Design
Thomas Kyte
McGraw-Hill Osborne Media, 2003

Optimizing Oracle Performance
Cary Millsap
O'Reilly, 2003

High Performance MySQL
Jeremy Zawodny and Derek Balling
O'Reilly, 2004

Microsoft SQL Server 2000 Performance Optimization and Tuning Handbook
Ken England
Digital Press, 2001

There are also many performance tuning courses available through one of the following training centers:

- Oracle University (*http://education.oracle.com*)
- Learning Tree International (*http://www.learningtree.com*)
- Microsoft Learning (*http://www.microsoft.com/learning*)
- MySQL Training (*http://www.mysql.com/training*)

Database Administration

Database administration is actually a multifaceted role that may encompass any or all of the following:

- Database server installation and configuration
- Database design
- Database programming, especially stored procedures, functions, and triggers
- Database security
- Backup and recovery
- Performance tuning

While larger organizations may employ one or more people specializing in each of the previous areas, smaller organizations often expect their database administrators to handle all of these tasks. If you will be performing design, programming, and tuning tasks, then please see the resource lists from earlier sections; all database administrators, however, should read a general-purpose administration book or take a training class so that they can become proficient in the core administration tasks, such as database installation and configuration, user creation and the assigning of privileges, backup and recovery strategies, and schema generation. Another essential resource for administrators is a SQL reference guide for your database server, which will show the syntax for the SQL schema statements, such as create index and alter table. Commonly used reference guides include:

> *Oracle Database 10g DBA Handbook*
> Kevin Loney, Bob Bryla
> McGraw-Hill Osborne Media, 2004

> *MySQL Administrator's Guide*
> MySQL Press, 2004

> *The SQL Server 2000 Book*
> Anthony Sequeira, Brian Alderman
> Paraglyph, 2003

There is also a plethora of administration training courses available from the following training centers:

- Oracle University (*http://education.oracle.com*)
- Learning Tree International (*http://www.learningtree.com*)
- Microsoft Learning (*http://www.microsoft.com/learning*)
- MySQL Training (*http://www.mysql.com/training*)

Report Generation

If you are charged with designing and/or generating reports for your organization, the two most important skills to acquire are:

- Knowledge of the capabilities of the reporting engine used by your organization
- Mastery of your database server's SQL implementation

While most of the reporting tools claim to generate SQL for you based on a visual representation of the report, I urge you to bypass this feature and craft your own SQL statements for all nontrivial reports. By doing so, you will know exactly what is being sent to the database server and will be better able to maintain and tune your reports over time.

Although some reporting engines are quite flexible regarding how the data on a report is sourced, many of the reporting tools require that all data for a single report be generated via a single query. Having a thorough understanding of SQL, especially subqueries (Chapter 9), set operations (Chapter 6), and conditional logic (Chapter 11) will allow you to produce far more sophisticated reports.

Some good books on reporting include:

Hitchhiker's Guide to SQL Server 2000 Reporting Services
Peter Blackburn, William Vaughn
Addison-Wesley, 2004

Business Objects: The Complete Reference
Cindi Howson
McGraw-Hill Osborne Media, 2003

Special Edition Using Crystal Reports 10
Neil FitzGerald et al.
Pearson Education, 2004

Index

Symbols

( ) (parentheses), condition evaluation, 57
+ (concatenation) operator, 116
' (single quotes), 107

A

access
    databases, 270
    transactions, 211–217
administration, databases, 276
advanced SQL, 268
aggregate functions, 136–142
    count(*), 137
    count( ), 137
aggregation, selective, 202
aliases
    columns, inserting, 41
    tables, defining, 47
all operator, 156
American National Standards Institute (see ANSI)
ANSI (American National Standards Institute), 6
    join syntax, 79
any operator, 158
applying
    indexes, 225
    mysql command-line tool, 14
    not operator, 58
    regular expressions, 68
    sets, 91–95
        operators, 95–100
    rules, 100–104

string data, 105–117
subqueries, 65, 165–174
    as tables, 83
temporal data, 122–124
wildcards, 67
Archive storage engine, 216
arguments
    greater than zero, 116
    single-argument numeric functions, 118
    truncate( ) function, 121
arithmetic operators, 117
ascending sort orders, 52
atomic clocks, 123
attaching check constraints, 24
autocommit mode, 213
auto-increment feature, starting, 28

B

bank schema, 34
BDB storage engine, 216
between operator, 62
bitmap indexes, 223
branch nodes, 222
bridges, 75
B-tree indexes, 222
building
    conditions, 59
    multicolumn indexes, 222
    schema statements, 24
    strings character by character, 109
built-in functions
    numeric, 118
    select statements, 40

We'd like to hear your suggestions for improving our indexes. Send email to *index@oreilly.com*.

G

generating
  constraints, 230
  data feed, 168
  dates, 128
  groups, 143–147
  numeric data, 117–122
  numeric key data, 28
  reports, 277
  strings, 106
  temporal data, 124–129
getutcdate( ) function, 123
GMT (Greenwich Mean Time), 123
granularities of locks, 211
Greenwich Mean Time (GMT), 123
group by clause, 49
grouping, 135–137
  aggregate functions, 137–142
  expressions, 144
  filtering, 147
  generating, 143–147
  implicit versus explicit groups, 138
  members, counting, 139
  multicolumn, 143
  rollups, 145
  single-column, 143
  subqueries, 169
guidelines for set operations, 94

H

having clause, 49
hierarchical database systems, 2
high-cardinality data, 224

I

implicit groups, 138
in operator, 153
indexes, 218–228
  applying, 225
  bitmap, 223
  B-tree, 222
  constraints, 229
  creating, 219
  deleting, 221
  full-text, 225
  modifying, 228
  multicolumn, 222
  nodes, 222
  show command, 220
  text, 225
  types of, 222
  unique, 221
  viewing, 220
inequality conditions, 60
inner joins, 74, 76
InnoDB storage engine, 216
insert statement, 28
  update statement, combining, 243
  values, 29
inserting
  column aliases, 41
  data into tables, 27
  indexes, 219
  intervals, 130
  keywords, 42
installing MySQL, 13
integers
  rounding, 120
  whole-number, 19
integration toolkits, 9
intermediate result sets, 82
intersect operators, 98
intersections, 92
intervals
  inserting, 130
  types, 129
into outfile clause, 240
invalid date conversions, 33
issuing queries without clauses, 15

J

joins, 74–81
  ANSI syntax, 79
  conditions, 88
  cross, 186–192
  equi-joins/non-equi-joins, 86–88
  natural, 192–194
  outer, 176–186
    conditional logic, 196
    left versus right, 180
    self, 183
    three-way, 182
  self-joins, 85
  three or more tables, 81–85

K

keyboards, special characters, 108
keys
  compound, 5
  foreign, 5
    constraints, 26, 228
    ER diagrams, 235

null values
   aggregate functions, 141
   case expressions, 207
   defining, 26
   filtering, 69–72
numbers
   aggregate functions, 142
   functions, returning, 132
   precision, controlling, 119
   rounding, 120
   string functions that return, 110
numeric built-in functions, 118
numeric data, generating, 117–122
numeric data types, 18–20
numeric key data, generating, 28
numeric placeholders, sorting via, 54

O

operations, sets, 91–95, 100–104
operators, 116
   all, 156
   any, 158
   arithmetic, 117
   between, 62
   concatenation (+), 116
   conditions, building, 59
   except, 99
   exists, 162
   in, 153
   intersect, 98
   not, applying, 58
   not in, 66
   regexp, 113
   set, 95–100
   union, 95
optimizers, 8
optimizing tables, 22
Oracle Text, 225
order by clause, 30, 50–54
   limit clause, combining, 238
ordered deletes/updates, 244–246
outer joins, 77, 176–186
   conditional logic, 196
   left versus right, 180
   self, 183
   three-way, 182
overlapping data, 95, 98

P

page locks, 211
parameters, limit clause, 239

parentheses ( ), condition evaluation, 57
passwords, mysql command-line tool, 14
physical models, 272
placeholders, sorting numeric, 54
PL/SQL language, 8
populating
   columns, generating strings, 106
   datetime columns, 126
   tables, 27–31
position( ) function, 111
positive values, signed data, 121
pow( ) function, 119
precedence
   numeric data, generating, 117
   set operations, 102
precision
   floating-point types, 19
   numbers, controlling, 119
primary keys, 5
   constraints, 24, 228
   nonunique, 32
procedural languages, 8
programming databases, 269

Q

queries
   clauses, 38–43
      from, 43–47
      group by, 49
      having, 49
      issuing without, 15
      order by, 50–54
      where, 47–49
   compound, 95
      set operation rules, 100–104
   conditional logic, 196
      case expression, 197–208
   cross joins, 186
   executing, 36–38
   grouping, 135–137
   indexes, 218–228
   joins, 74–81
      conditions, 88
      equi-joins/non-equi-joins, 86–88
      self-joins, 85
      three or more tables, 81–85
   limit clause, 237
   multiuser databases, 210
   natural joins, 192–194
   outer joins, 179
   ranking, 240
   versioning, 211

sign( ) function, 122
signed data, 121
simple case expressions, 200
single quotes ('), 107
single-argument numeric functions, 118
single-column
    grouping, 143
    subqueries, 153
single-parent hierarchies, 2
sizing text types, 17
soft errors, 33
solutions to exercises, 250–266
sorting
    ascending/descending sort orders, 52
    collation, 64
    compound query results, 101
    expressions, 53
    numeric placeholders, 54
SQL, defining, 6–11
SQL92 join syntax, 79
starting
    auto-increment feature, 28
    Configuration Wizard, 13
    transactions, 213
statements
    classes, 7
    containing, 150
    create table, 25
    insert, 28
        values, 29
    overview of, x
    schemas, building, 24
    scope, 150
    select
        extensions, 237–243
        query clauses, 38–43
    subqueries, 150
        applying, 165–174
        correlated, 160–164
        noncorrelated, 152–160
        types of, 151
    troubleshooting, 32
storage engines, 211
    selecting, 216
strcmp( ) functions, 112
strings
    applying, 105–117
    built-in functions, 110
    converting, 133
    escaping, 107

functions
    returning, 114, 131
        that return numbers, 110
        that return strings, 114
    generating, 106
    manipulating, 110
    ranges, 63
    string-to-date conversions, 127
    temporal data, generating, 124
    truncating, 106
    wildcards, applying, 67
str_to_date( ) function, 128
stuff( ) function, 117
subqueries
    applying, 65, 165–174
    as data sources, 165
    as expression generators, 170
    in filter conditions, 170
    grouping, 169
    multicolumn, 159
    multiple-row, 153
    overview of, 150
    scalar, 152
    single-column, 153
    subquery-generated tables, 44
    tables, applying as, 83
    task-oriented, 168
    types of, 151
        correlated, 160–164
        noncorrelated, 152–160
syntax
    ANSI join, 79
    case expressions, 198

T

tables, 4, 6, 210
    aliases, defining, 47
    constraints, 228–233
    creating, 22–27
    cross joins, 186–192
    deleting, 31
    design, 22
    ER diagrams, 235
    fabrication, 166
    from clause, 43–47
    group by clause, 49
    having clause, 49
    indexes, 218–228
    links, 46
    locks, 211

users
  multiuser databases, 210
  transactions, 211–217
UTC (coordinated universal time), 123

V

values
  columns, troubleshooting, 33
  converting, 133
  distinct, counting, 139
  insert statements, 29
  null
    aggregate functions, 141
    case expressions, 207
    filtering, 69–72
  signed data, 121
Varchar data type, 105
versioning, 211

viewing indexes, 220
views, queries, 45

W

where clause, 47–49
  conditions, evaluating, 56–59
  group filter conditions, 147
whole-number integers, 19
wildcards, applying, 67
Windows, installing MySQL, 13
wizards, starting Configuration Wizard, 13
write locks, 210

Z

zeros, 19
  arguments greater than, 116
  division by zero errors, 205

About the Author

Alan Beaulieu has been designing, building, and implementing custom database applications for over 15 years. He specializes in designing Oracle databases and supporting services in the fields of financial services and telecommunications. He is also an expert with the SQL language and specializes in report design and implementation across multiple database servers. Alan has a bachelor of science degree in operations research from the Cornell University School of Engineering. He lives in Massachusetts with his wife and two daughters and can be reached at *learning_sql@yahoo.com*.

Colophon

Our look is the result of reader comments, our own experimentation, and feedback from distribution channels. Distinctive covers complement our distinctive approach to technical topics, breathing personality and life into potentially dry subjects.

The animal on the cover of *Learning SQL* is an Andean marsupial tree frog (*Gastrotheca riobambae*). As its name suggests, this crepuscular and nocturnal frog is native to the western slopes of the Andes mountains and is widely distributed from the Riobamba basin to Ibarra in the north.

During courtship, the male calls ("wraaack-ack-ack") to attract a female. If a gravid female is attracted to him, he climbs onto her back and performs a common frog mating hold called the nuptial amplexus. As the eggs emerge from the female's cloaca, the male catches the eggs with his feet and fertilizes them while maneuvering them into a pouch on the female's back. A female may incubate an average of 130 eggs, and development in the pouch lasts between 60 and 120 days. During incubation, swelling becomes visible, and lumps appear beneath the skin on the female's back. When the tadpoles emerge from the pouch, the female tree frog deposits them into the water. Within two or three months the tadpoles metamorphose into froglets, and at seven months they are ready to mate ("wraaaack-ack-ack").

Both the male and female tree frog have expanded digital discs on their fingers and toes that help them climb vertical surfaces such as trees. Adult males reach 2 inches in length, while females reach 2.5 inches. Sometimes they are green, sometimes brown, and sometimes a combination of green and brown. The color of the juveniles may change from brown to green as they grow.

Matt Hutchinson was the production editor for *Learning SQL*. Octal Publishing, Inc. provided production services. Adam Witwer, Jamie Peppard, Genevieve d'Entremont, and Claire Cloutier provided quality control.

Ellie Volckhausen designed the cover of this book. The cover image is from the Dover Pictorial Archive. Karen Montgomery produced the cover layout with Adobe InDesign CS using Adobe's ITC Garamond font.

David Futato designed the interior layout. This book was converted by Keith Fahlgren to FrameMaker 5.5.6 with a format conversion tool created by Erik Ray, Jason McIntosh, Neil Walls, and Mike Sierra that uses Perl and XML technologies. The text font is Linotype Birka; the heading font is Adobe Myriad Condensed; and the code font is LucasFont's TheSans Mono Condensed. The illustrations that appear in the book were produced by Robert Romano, Jessamyn Read, and Lesley Borash using Macromedia FreeHand MX and Adobe Photoshop CS. The tip and warning icons were drawn by Christopher Bing. This colophon was written by Lydia Onofrei.

Better than e-books

Buy *Learning SQL* and access the digital edition FREE on Safari for 45 days.

Go to www.oreilly.com/go/safarienabled
and type in coupon code GDKS-ZIDJ-RS27-B7J9-WIRQ

Search
thousands of
top tech books

Download
whole chapters

Cut and Paste
code examples

Find
answers fast

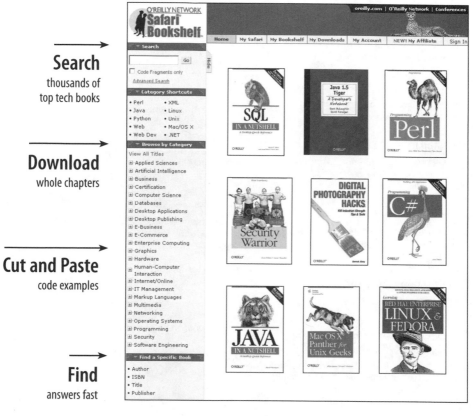

Search Safari! The premier electronic reference
library for programmers and IT professionals.

Keep in touch with O'Reilly

Download examples from our books

To find example files from a book, go to: *www.oreilly.com/catalog* select the book, and follow the "Examples" link.

Register your O'Reilly books

Register your book at *register.oreilly.com* Why register your books? Once you've registered your O'Reilly books you can:

- Win O'Reilly books, T-shirts or discount coupons in our monthly drawing.
- Get special offers available only to registered O'Reilly customers.
- Get catalogs announcing new books (US and UK only).
- Get email notification of new editions of the O'Reilly books you own.

Join our email lists

Sign up to get topic-specific email announcements of new books and conferences, special offers, and O'Reilly Network technology newsletters at:

elists.oreilly.com

It's easy to customize your free elists subscription so you'll get exactly the O'Reilly news you want.

Get the latest news, tips, and tools

www.oreilly.com

- "Top 100 Sites on the Web"—PC Magazine
- CIO Magazine's Web Business 50 Awards

Our web site contains a library of comprehensive product information (including book excerpts and tables of contents), downloadable software, background articles, interviews with technology leaders, links to relevant sites, book cover art, and more.

Work for O'Reilly

Check out our web site for current employment opportunities:

jobs.oreilly.com

Contact us

O'Reilly Media, Inc.
1005 Gravenstein Hwy North
Sebastopol, CA 95472 USA
Tel: 707-827-7000 or 800-998-9938
    (6am to 5pm PST)
Fax: 707-829-0104

Contact us by email

For answers to problems regarding your order or our products:
order@oreilly.com

To request a copy of our latest catalog:
catalog@oreilly.com

For book content technical questions or corrections: **booktech@oreilly.com**

For educational, library, government, and corporate sales: **corporate@oreilly.com**

To submit new book proposals to our editors and product managers:
proposals@oreilly.com

For information about our international distributors or translation queries:
international@oreilly.com

For information about academic use of O'Reilly books:
adoption@oreilly.com
or visit:
academic.oreilly.com

For a list of our distributors outside of North America check out:
international.oreilly.com/distributors.html

Order a book online

www.oreilly.com/order_new

O'REILLY®